BAMBOOS OF NEPAL

AN ILLUSTRATED GUIDE

Chris Stapleton

Illustrations of the genera and species, with notes on
identification, distribution, utilisation, and propagation

BAMBOOS OF NEPAL:
AN ILLUSTRATED GUIDE

Chris Stapleton

Forestry Department, University of Aberdeen
Royal Botanic Garden Edinburgh
Royal Botanic Gardens Kew

in association with

Forestry Research and Information Centre
Department of Forestry and Plant Research
His Majesty's Government of Nepal
Kathmandu

Royal Botanic Gardens, Kew, on behalf of

The Overseas Development Administration, London

Forestry Research Programme, University of Oxford

Published by The Royal Botanic Gardens, Kew

for The Overseas Development Administration of the British Government
Forestry Research Programme
University of Oxford,
Halifax House,
6 South Parks Road,
Oxford OX1 3UB

First published 1994

Design, illustrations, and layout by the author. Cover by Media Resources, RBG Kew.

Research for this guide and its production were funded by the Overseas Development Administration, under research grants R4195 and R4849. Field work was implemented by the Forestry Department of Aberdeen University in conjunction with the Department of Forestry and Plant Research of His Majesty's Government of Nepal. Illustrations and camera-ready copy were produced at the Royal Botanic Garden Edinburgh. Final editing and production were supported by the Anglo-Hong Kong Trust.

ISBN 0 947643 68 0

Printed in Great Britain by Whitstable Litho Ltd.

CONTENTS

INTRODUCTION

BAMBOOS are widely distributed throughout Nepal, but they are more common in the eastern half of the country, from Dhaulagiri to the Sikkim border. In higher rainfall areas such as those around Pokhara and Ilam, a wider variety of genera and species can be found, as well as larger numbers of bamboo clumps. Temperate and sub-alpine genera which are more common in Tibet and Bhutan can be found at altitudes of up to 4,000m in eastern Nepal. Tropical species from Malaysia and Burma extend into the Nepalese terai.

Bamboos are widely planted on private land, and they are also important minor forest products, several species being systematically harvested on an annual basis. While traditional uses continue to use large quantities of bamboo, new uses are also being developed, and substantial export markets remain to be explored.

There are very many uses for bamboo, and it is treated as a multipurpose raw material from which almost anything can be made. Because bamboo is easy to split, even large culms can be converted into usable sections without anything more sophisticated than a khukri. When used as whole sections, bamboo pillars are extremely strong for their weight, and sections of some species can be quite hard. Most bamboos, however, are flexible and not very durable, and they will be attacked by fungus and insects faster than timber from trees, so that they need to be replaced on a regular basis, or adequately preserved. The flexibility of softer species allows the weaving of thin strips into all manner of baskets and trays, used for collecting, sorting, storing and transporting food crops and other products.

The role of bamboos in soil conservation is very important in the Himalayas. Because of their dense surface roots, bamboos can provide good protection against sheet and gully erosion. The large mass of the rhizome system can form an effective buttress, holding up terraces and road banks. In combination with those tree species which root to a greater depth, bamboos are now an important component of bio-engineering techniques in Nepal, providing a low cost means of slope stabilisation as well as useful products.

Bamboos are harvested by thinning older poles. The rest of the clump will continue to protect the soil, and will produce new culms without the need for replanting. The new shoots appear at a time when most animals are not allowed to graze uncontrolled, in order to protect field crops. They are also well protected by tough sheaths with irritant hairs, and they grow rapidly, so that the tender shoot tips are soon out of the reach of grazing animals. In this way the bamboo growth cycle is well co-ordinated with animal husbandry in the middle hills once clumps have become established, and bamboos will continue to provide a sustained yield of poles and fodder on an annual basis for considerable lengths of time.

The taxonomy of bamboos is quite complicated, and it has been neglected for a long time. Bamboos are giant grasses, but they differ from the smaller grasses in many ways. They have woody culms, well developed branching, specialised culm sheaths, leaf bases narrowed into thin petioles, and cyclical flowering. For a long time taxonomists thought that the flowers were essential for identification, but now it is accepted that vegetative parts are also important. As some species may wait up to 150 years before flowering this makes it much easier to identify the different species.

There has recently been a period of great confusion over the genera of small Himalayan bamboos. The species were originally all placed in the genus *Arundinaria*. Over the last century, and during the 1980's in particular, many new genera were described in Japan and China. These genera were not always clearly defined, and so were often not recognised in other countries. Now that much more information on Sino-Himalayan bamboos is available to the scientific world many of these new genera are becoming more widely accepted. However certain new genera such as *Sinarundinaria* have been shown to be the same as genera which had already been named, so they are rejected. A more stable system of genera is now recognised, and is it hoped that it will not be necessary to make many more changes to generic names.

Species are described in this guide to allow positive identification in the field from vegetative material alone. It is aimed at forestry and agricultural personnel rather than specialised taxonomists, therefore the terminology is kept as simple as possible, and important features of each genus and each species are illustrated. Accurate identification of species requires a little detailed knowledge of the parts of bamboo plants. Therefore the most important parts of the bamboo plant are briefly described. The genera are separated using a key which does not require flowers, and a full glossary is given. The important characters of each genus are illustrated. Individual species are then described, showing how to distinguish them from closely-related species, and giving some basic information on distribution, uses, and appropriate propagation techniques. This information is not comprehensive, but it is hoped that once the different species can be recognised in the field more accurately, it will be possible for others to gather more useful information on them. Garden species grown exclusively as ornamentals in Kathmandu, such as *Phyllostachys nigra* and *P. pubescens* are not included.

Many other species undoubtedly remain undiscovered in less accessible parts of the country. Western Nepal beyond Palpa district, areas of the terai close to the Indian border, and most of the temperate forest areas are not covered. Hopefully most of the more common species of the middle hills of central and eastern Nepal have been included, and this guide can provide the basis for further studies.

No account of bamboos would be complete without some reference to their peculiar flowering behaviour. Scientists still do not know how these plants manage to mark the passage of time, so that they can flower synchronously after an interval of up to a century or more. It is now known, however, that species have different flowering habits. Lengths of flowering cycle vary greatly between varieties and species, and not all species will die after flowering. Because identification has been inadequate in the past, it is not yet possible to list the flowering habits of each species, but hopefully this guide will allow better records to be kept in the future.

IDENTIFYING BAMBOOS

To identify bamboo species the most important parts of the plant are the culm sheaths. These are protective sheaths around the stems (the stem is called a culm in all grasses), see fig. 1. The sheaths below the leaves (leaf sheaths) are also important, see fig. 2). At the top of these sheaths there is a projecting tongue in the centre called the ligule, and ears on each side called auricles. The shape and size of the

Culm sheaths at the culm base are different from those higher up. They are broader and have shorter blades. To standardise descriptions, culm sheaths at eye-level on the large bamboos are taken. These are approximately ¼ of the way up the culm. Smaller bamboos are treated in the same manner, culm sheaths from ¼ of the way up the culm from the base being described.

New culm sheaths show the features of the species best. Older sheaths often have parts that are missing or have rotted away, especially in hotter areas,

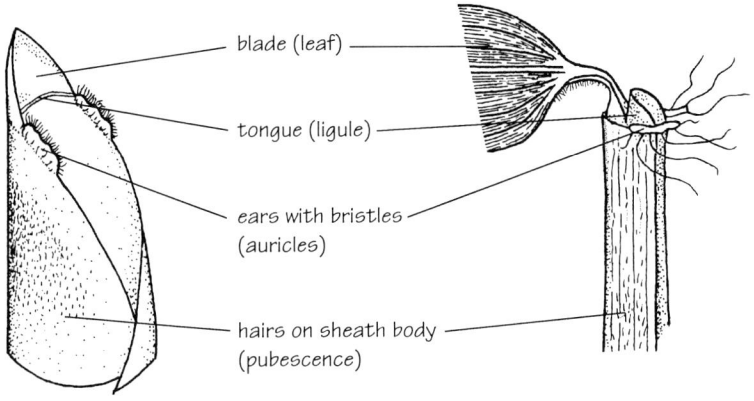

blade (leaf)

tongue (ligule)

ears with bristles
(auricles)

hairs on sheath body
(pubescence)

fig. 1 - culm sheath   fig. 2 - leaf sheath

auricles, and whether there are stiff bristles on their edges are all important. The shape, length and the type of edge on the ligule are also important.

The blade of the culm sheath is a modified leaf. Its shape, whether it has hairs on the back or around the base, whether it is erect or bent backwards (reflexed), and whether it falls off early (deciduous), or will remain attached (persistent), are also all important.

and for this reason bamboos are easiest to identify in the late summer and autumn. In winter and spring care must be taken to find undamaged sheaths. In the same way, leaf sheaths are damaged by strong wind and rain so that auricles, bristles, and hairs are blown off after a few months. Small new leaves can be found at most times of year except in winter. Some of these should be collected as well as larger, older leaves. The drawings in this guide show fairly new parts. These are typical of those which would be found in October.

The surface of the culm is also important. Young culms have a coating of wax, which can be either thick and furry, or thin, and either light or dark in colour, and it may rub off quickly to leave the culms shiny, or it may persist so that the culms stay matt and dull. The joints of the culm (nodes) may be raised or level, with rings of different colours, and they may bear small aerial roots or thorns. The surface of the culm may be rough with tiny sharp points, or smooth, or it may be covered in small vertical ridges.

Branching is a very important characteristic in bamboos, especially in the separation of genera. The number of branches in the first year of growth is important, as well as the eventual number of branches which older culms develop. Whether these branches are all the same size should be noted, or whether the central branch is much larger than all the others, in which case it may develop aerial roots on its base.

Rhizomes are difficult to examine as they usually remain under the ground. The type of rhizome will determine whether the bamboo grows in a clump (clump-forming), or spreads widely (running). Running rhizomes may have roots at all nodes, or they may have roots only on the short internodes near the bases of the culms, and they may be solid or hollow.

Flowers of bamboos are occasionally found. There are two different kinds of bamboo inflorescence. One type will keep on branching to give dense clusters or rounded balls of flowers, which are well-developed in genera *Dendrocalamus,* and *Bambusa,* and in *Cephalostachyum.*

The other type produces flowers in large panicles, similar to those of an ordinary grass. The panicles and flowers of *Thamnocalamus* remain partially hidden by sheaths, while the sheaths fall quickly from the panicles of other genera such as *Drepanostachyum* and *Yushania.*

The colour of the flowers can allow quick identification of large bamboos if the flowers are young, but they all fade to a brown or straw colour before long. *Dendrocalamus hamiltonii* var. *hamiltonii* has purple flowers with distinctive red anthers. *D. hookeri* has olive-green to brown flowers. *Bambusa tulda* and *B. nutans* have green flowers, while *B. balcooa* and *B. nepalensis* have green flowers with purple tips. *D. giganteus* has very long pendulous sprays of flowers.

To allow accurate identification of bamboos in the herbarium, a collection of leaves and culm sheaths is usually adequate if they are in good condition and well protected. If a proper plant press is not available, the leaves can be packed inside a rolled culm sheath, and a series of culm sheaths can be rolled together and tied firmly. The outer sheaths will protect the delicate parts such as auricles and blades of those at the centre. For the small bamboos a section of the culm is very useful, including a node with its branches cut back to 5cm. For spreading bamboos a short section of the rhizome is also required. This can often be found on an over-hanging bank or road-cutting. If flowers are found, leaves and culm sheaths should always be searched for and included if possible, even if they are old. However, if they come from a different clump this should be made clear. Collections should never be put into plastic bags as they will rapidly go mouldy.

PROPAGATION METHODS

H IMALAYAN bamboos can be propagated either from seed, or by vegetative methods. Seed should always be used when a suitable species or variety is flowering, as long as there is a nursery in which the seedlings can be grown. This guarantees the maximum period of vegetative growth before flowering begins on the planted material. Various techniques of vegetative propagation can also be used when seed is not available. The traditional propagation technique, which is essentially clump division, and involves digging out a section of rhizome, has to be used when nursery facilities are not available. Other kinds of cuttings can be used when there is a reliable nursery nearby. Different forms of propagation or type of cutting are appropriate for different genera and species. Plants raised in nurseries either from seed or from cuttings can be produced in large quantities, but being smaller than the traditional planting material, they require better protection from grazing animals.

RAISING SEEDLINGS

Collection of seed from flowering clumps is best organised by local private seed collectors. It requires local knowledge of where and when bamboos are flowering, and a rapid response to obtain good quality seed before it is destroyed by insects, rain or fire. Seed of *Bambusa* and *Dendrocalamus* species may be produced within three weeks of the start of flowering. Unlike agricultural grasses, which have been bred to retain their seed, the seed of bamboos will often fall to the ground as soon as it is mature. Collection of good seed involves collection of the seed as it falls by placing sheets or tarpaulins underneath the flowering clumps, or collection of the fallen seed from the litter and vegetation on the ground. Collection of the flowering branches usually results in loss of most of the viable seed as it is so easily dislodged, although some temperate bamboos retain their seed in the flowers for longer than the subtropical bamboos.

The seed should be dried in the sun and cleaned. The chaff can be separated from the seed by gentle rubbing and winnowing. Insects may destroy the seed completely within a few months if they are not eliminated. The principal pest is *Sitotroga cerealella*, a small light brown moth with tiny larvae that burrow into the seed. They eat the seed contents, leaving white papery remnants of their cocoons, and they can complete their life cycle in 5 weeks. Treatment with insecticide powder or placing the dried seed in a freezer for 3 days is necessary to control this pest.

Storage of bamboo seed is very difficult, even after elimination of insect pests. It can be dried, but even when the moisture content is reduced to the ideal level of around 8-10%, and the temperature is maintained at 5°c, the germination rates will still fall to 25% or less in the first year. This means that some seed may be stored to be sown in a second season, but most of it should really be sown before the first monsoon. Longer periods of storage are possible if dried seed could be stored at a constant -18°c, but repeated thawing and freezing is likely to kill the seed.

Seed of large subtropical bamboos such as *Dendrocalamus hamiltonii* has no dormancy and fresh seed will normally germinate within two weeks if conditions of temperature and humidity are suitable. Seed of the smaller subtropical and temperate bamboos may have substantial dormancy, and it might germinate more quickly after a period of cold pre-treatment, such as stratification or refrigeration at 5°c. *Himalayacalamus hookerianus* seed stored and sown at 20-25°c. germinated very slowly over a nine month period, the first shoot appearing on the 35th day after sowing.

Seed can be sown into seedbeds or it can be sown directly into containers. Seedlings of *Dendrocalamus strictus* can be grown without shading. The seedlings of most of the Himalayan bamboos require good shading and frequent watering, and they may be difficult to transplant without loss.

As it is not always possible to collect the seed in time, it may sometimes be necessary to collect the germinated seedlings from under flowering clumps in the monsoon after flowering. Very dense regeneration is sometimes seen, if the pressure from grazing animals is low and there are no fires to destroy the seed. These natural seedlings can be transplanted directly into containers in shaded beds, but they need to be kept moist during transportation, and they will lose their leaves or die if they are not transplanted promptly.

TRADITIONAL PROPAGATION

Only one technique of propagation has been used on any scale in the Himalayas in the past. An accessible culm from near the edge of the clump is removed by digging around its rhizome, cutting the rhizome where it branches from its mother. The length of the culm

is reduced by cutting it 1-2m above the ground. This propagation technique is undertaken at the start of the monsoon. If undertaken too early the roots will dry out and the rhizome will die. If left too late buds on the rhizome will have already developed into fragile new shoots, which will die during the transplanting process. If the operation is successful, the culm will grow new

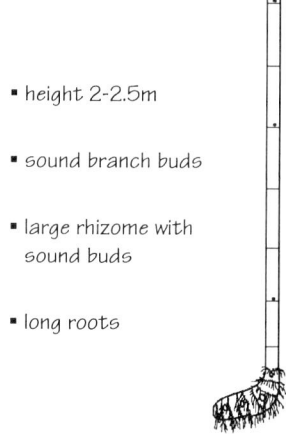

- height 2-2.5m

- sound branch buds

- large rhizome with sound buds

- long roots

fig. 3 -traditional planting material

branches and leaves at the top in the first year, and in the second year a new shoot will emerge from the rhizome, hopefully reaching a height of several metres. Small bamboo species should be planted using several culms and rhizomes still joined together.

This is a robust planting method, which can establish bamboo clumps in areas where there is substantial grazing of livestock. The drawbacks are the high labour costs, and the shortage of planting material available, which both limit the scale upon which it can be implemented. However, this is probably the only technique which will be successful in planting areas where grazing cannot be prevented.

It is possible to improve the success rates and the speed of establishment achieved using the traditional technique. Selection of older culms with more reserves in their rhizomes will provide better tolerance to drought and grazing. However, older rhizomes are more difficult to extract. Use of a longer culm section of 2-2.5m reduces the browsing of the new growth at the top of the culm, but this makes transport more difficult. Support of the culm with two poles in a tripod arrangement will prevent animals from pushing it over to reach the leaves. Watering of the roots during extraction, transport, and during periods of drought will greatly improve growth rates. Protection of new shoots and foliage with branches from thorny bushes may reduce the damage caused by animals.

CULM CUTTINGS

Many species of *Dendrocalamus* and *Bambusa* produce aerial roots from the bases of larger branches. The rooting of branches can be used to raise new plants without the extraction of rhizomes. Various propagation techniques based upon the rooting of branches are used in areas of the tropics with heavy spring rainfall. They are not commonly used in monsoonal areas with a spring drought. This is because branches grow in the spring, and without moisture they cannot root effectively. Where nursery facilities are now available to provide abundant artificial watering from spring until the monsoon, culm cuttings of many species can be highly successful. As well as frequent and regular watering,

cuttings also require good shading and protection from grazing animals.

Branches on their own rarely have sufficient reserves to sustain strong new shoots until they root. Sections of the culm with branch bases attached are more successful. The principal limiting factor in the growth of new shoots is usually water availability. Single-node culm sections planted horizontally with both ends buried expose a large area of vascular culm tissue to the wet soil. This maximises the entry of water first into the culm section, and then into each branch and its buds.

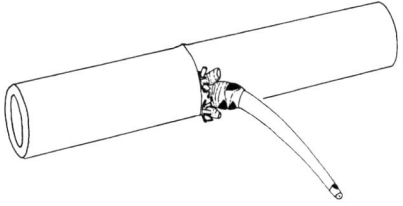

fig. 4 - single-node culm cutting

Cuttings are taken just before growth of new branch shoots and leaves in the spring, normally from mid-March to mid-April. Large 2-year-old culms with strong branches should be chosen. These culms would often be harvested during the previous winter, so it is advisable to buy and mark them earlier. The buds at the base of the central branches must not be damaged. The culms are felled and the branches are trimmed back. The central branch is cut at a length of about 20cm, beyond the first long internode, while the smaller branches are cut right back to the culm. The culm is then cut into single-node sections, each one bearing a strong branch or a dormant bud. The cuttings are then covered with wet sacking, and transported to the nursery.

Nursery beds are best prepared from soil that has been cultivated for many years and is free from cockchafer larvae, which could quickly destroy the new roots. Heavy soils are preferable as they will retain water more effectively. Shading must be provided over the beds. Termites should be eliminated.

Cuttings are set in the soil so that the culm ends and the branch bases are just below the soil level. If the branch base has more buds on one side than the other, the side with most buds should face downwards. Downward-facing buds are more likely to give rooted shoots.

The culm cuttings have enough reserves to support shoot growth for 2-3 months. After that the shoots will start to die back, but roots should just be beginning to develop from the bases of the new shoots. Much larger shoots will then arise from the cutting beds. This second generation of shoots will have abundant rooting, and will form the planting material, once it has hardened off and developed its own branches and leaves. The time required for production of reliable planting material varies from 6 months to 2 years. Most plants will be suitable for planting in the second monsoon after the cuttings were taken. After lifting, the plants should be kept in a shaded area, and watered occasionally until they are planted.

The use of rooting hormones to improve the success of culm cuttings is sometimes advocated. The benefits have not yet been proven in a statistically valid trial, and other factors such as selection of sound material, timely setting of cuttings, and maintaining good environmental conditions are probably more important in most cases.

OTHER PROPAGATION TECHNIQUES

A technique recommended in China uses whole culms buried horizontally, with their rhizomes still attached. This technique produces rooting plants all along the culm in China. Each plant develops from shoots growing from branch buds. In a trial of this technique in Nepal very few plants were produced, even with frequent watering above the rhizome, and with notches cut above each node. This is probably because of the Himalayan spring drought. Branch buds stayed dormant for several years, and shoots only grew from the rhizomes.

A similar technique has sometimes been recommended in India. Holes are cut into each internode of a horizontally buried culm, and the internodes are filled with water. Although this may seem to be a good idea, the interior lining of bamboo internodes is almost entirely waterproof, so this water cannot be transported to the buds and shoots.

In *Bambusa multiplex* small rooting rhizomes are often produced in the air from branch bases. These offsets will develop into successful plants if removed and planted, but only during the monsoon. Similar offsets are produced on *Bambusa nutans* subsp. *cupulata,* but attempts to propagate from them have only been made in the spring, and they were not successful.

Tissue culture has not yet been successful without seed to produce embryogenic callus. Once callus is available however, planting material can be produced indefinitely. Transporting plants from a central laboratory to planting sites is feasible in the terai.

9

FIELD KEY TO BAMBOO GENERA FOR USE IN NEPAL

Culm sheath blade (from lower ½ of culm) broad, length less than twice the width:-

Culm covered with dark or thick fur, central branches
varied, often very large .... 1. *Dendrocalamus*
Culm with light covering of pale wax, central branches
fairly uniform, usually quite small .... 2. *Bambusa*

Culm sheath blade (from lower ½ of culm) narrow, length more than twice width:-

Clump-forming bamboos, culms growing in separate
clumps of more than 10 culms:-

    Leaves with short cross veins as well as long veins; buds tall, chilli-shaped:-

        Culms erect and culm surface smooth .... 3. *Thamnocalamus*
        Culms curving outwards strongly at the base,
        or culm surface with fine ridges .... 4. *Borinda*

    Leaves with only long veins, no cross veins; buds short, onion-shaped:-

        Culm nodes with projecting wavy corky collar; culm sheath
        edges with long comb-like fringe .... 5. *Ampelocalamus*
        Culm nodes with flat even corky collar; culm sheath
        edges without long comb-like fringe:-

            Internodes long, up to 50cm .... 6. *Cephalostachyum*
            Internodes short, up to 30cm:-

                Culm sheath interior rough below ligule;
                branches 20 - 70 .... 7. *Drepanostachyum*
                Culm sheath interior smooth below ligule;
                branches 10 - 30 .... 8. *Himalayacalamus*

Spreading bamboos, culms growing separately or in groups of up to 10 culms:-

Leaves with no cross veins, long veins only .... 9. *Melocanna*
Leaves with distinct cross veins as well as long veins :-

    Long rhizome lengths without roots .... 10. *Yushania*
    Rhizome rooting at all nodes .... 11. *Arundinaria*

1. DENDROCALAMUS

A genus containing the largest of all bamboo species, forming clumps up to 30m tall. The culms are thin-walled and covered with thick furry wax when young. The branches are usually absent lower down the culm, and are very variable in size, some being more than 5cm in diameter. The bracts at the base of each spherical inflorescence have one ciliate keel (fig. 8), in contrast to those of *Bambusa*, which have two ciliate keels (fig. 12). Most of the species are from subtropical to warm temperate areas, withstanding only a few degrees of frost, although one tropical Malaysian species is planted in the terai. All species are easy to propagate by culm cuttings, as the large branches readily produce roots. The thin walls of the culms make the young shoots more liable to attack by shoot-boring larvae, and the dried poles are readily attacked by beetles if not preserved. This genus contains the most important species for edible shoot production in the Himalayas, as well as several general multipurpose species.

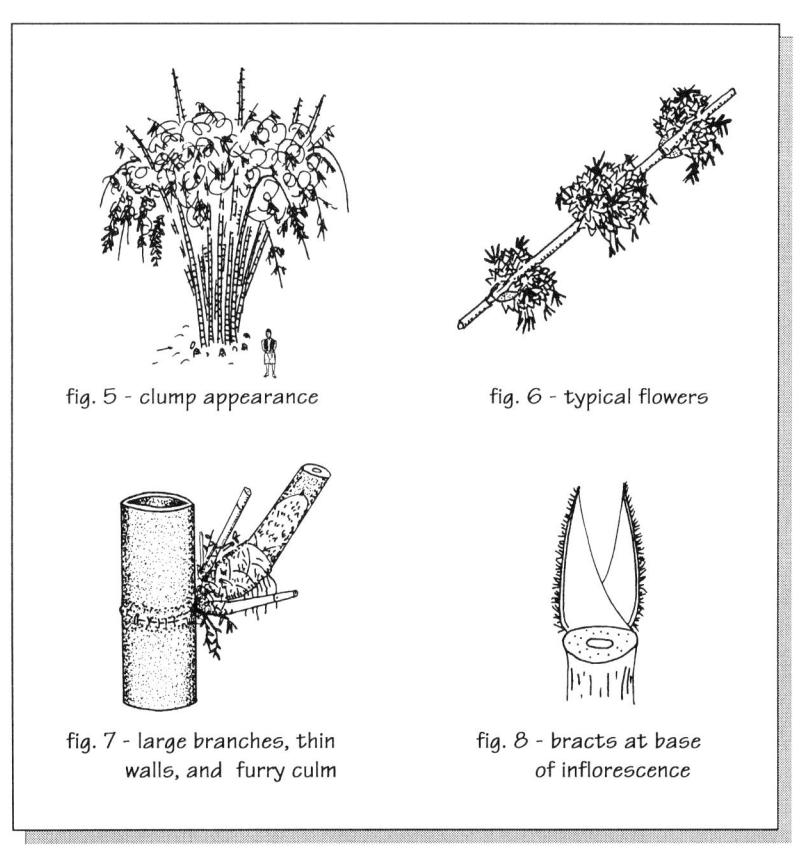

fig. 5 - clump appearance

fig. 6 - typical flowers

fig. 7 - large branches, thin walls, and furry culm

fig. 8 - bracts at base of inflorescence

KEY TO *DENDROCALAMUS* SPECIES

Culm sheath auricles more than 2cm wide . . . . *giganteus*
Culm sheath auricles less than 2cm wide:-

    auricles small and round with bristles:-

        culm and sheath with dark brown fur or hairs . . . . *hookeri*
        culm and sheath with light fur or hairs . . . . see *Bambusa nepalensis*

    auricles absent, or small, triangular, and naked:-

        culms thick-walled or solid, leaf sheaths with no hairs . . . . *strictus*
        culms thin-walled, leaf sheaths with hairs at first

        culms remaining dull with persistent fur,
        leaf sheaths with white hairs, branchlets not thorny:-

            culm internodes uniformly cylindrical   . *hamiltonii var. hamiltonii*
            culm internodes irregularly swollen . . . . *hamiltonii var. undulatus*

        culms becoming shiny, fur deciduous,
        leaf sheaths with brown hairs, branchlets thorny,   see *Bambusa balcooa*

Dendrocalamus giganteus   (Nep. *dhungre bans, rachhasi bans*)   D30

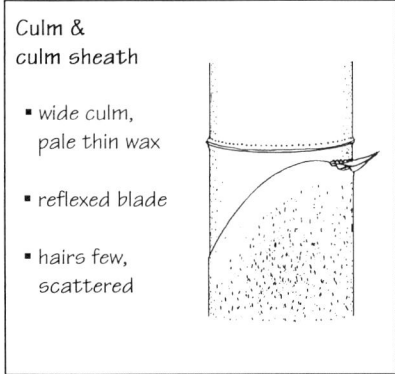

Culm &
culm sheath

- wide culm,
  pale thin wax

- reflexed blade

- hairs few,
  scattered

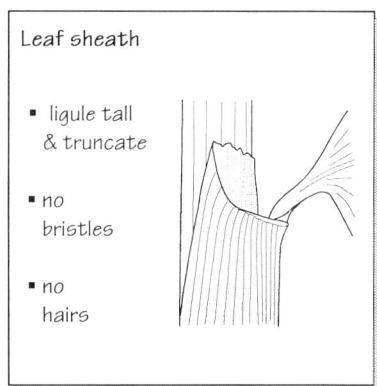

Leaf sheath

- ligule tall
  & truncate

- no
  bristles

- no
  hairs

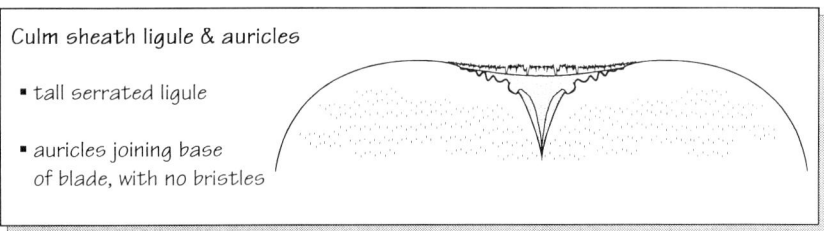

Culm sheath ligule & auricles

- tall serrated ligule

- auricles joining base
  of blade, with no bristles

This is the largest bamboo in South Asia, with a maximum diameter of more than 30cm. The tallest culms reach 30m in height. It is similar to the slightly smaller *B. balcooa,* both species having no bristles at the top of the culm sheath. However, this species has horizontal culm sheath blades, and it can also be distinguished from *B. balcooa* by the absence of hairs on the leaf sheaths. The hairs on the culm sheath are also much fewer, lighter in colour, and are flattened against the sheath. The glabrous culm sheath auricles can distinguish it from *Dendrocalamus hookeri.*

The leaf sheaths become quite red at the tips, and it has long pendulous flowering branches. The large diameter culms are used as pillars or for making storage containers, and for special uses such as road barriers. However, they are too large for most general purposes, and this species is not widely cultivated. The very large leaves are used as animal fodder.

This species is found across the plains of West Bengal and Assam, and has been planted in the eastern terai. It is a tropical species from Malaysia, and may not grow well above 1,000m.

Propagation of this species is not easy. The large size of the rhizomes makes it difficult to use the traditional technique. Culm cuttings would be successful as the branches are large, but there are few branches in the lower part of the culm.

Dendrocalamus hamiltonii　(Nep. *choya, tama, ban bans*)　　　D4/D6
var. **hamiltonii**　and var. **undulatus**

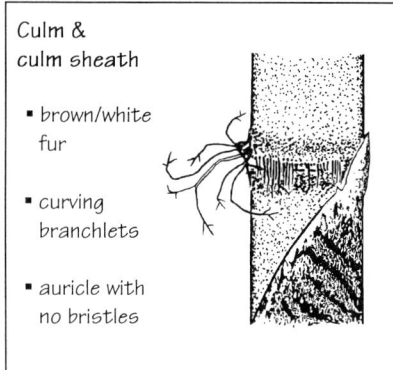

Culm &
culm sheath

▪ brown/white
fur

▪ curving
branchlets

▪ auricle with
no bristles

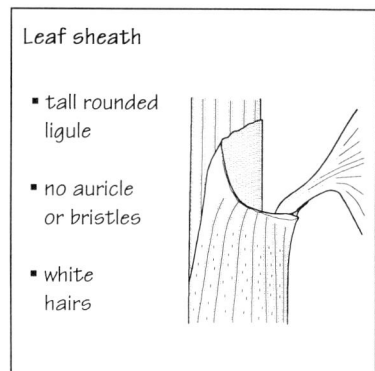

Leaf sheath

▪ tall rounded
ligule

▪ no auricle
or bristles

▪ white
hairs

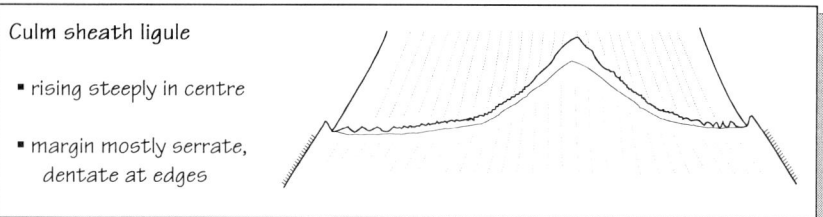

Culm sheath ligule

▪ rising steeply in centre

▪ margin mostly serrate,
dentate at edges

The most common bamboo of the subtropical forest along the outermost foothills of the entire Himalayan range, often cultivated further into the hills.

D. hamiltonii has a long leaf sheath ligule, naked triangular auricles on the culm sheaths, persistent pale fur on the culms and long drooping culm tips. There are several varieties of this species, differing in straightness of culm and degree of branching. Var. *hamiltonii* has straight culms, heavy branching and red anthers. Var. *undulatus* has shorter swollen culm internodes, more dimpled culm sheath blades, and yellow anthers.

The culms are thin-walled and very flexible, giving the best weaving material of all large bamboos, but the large branches of many varieties make the culms difficult to split. This species is commonly managed without cutting mature culms. New shoots are removed for human consumption, and the large branches are cut for weaving material, and for fodder. This often leads to tightly congested clumps.

Propagation by vegetative means is easy because of the large branches and prolific aerial rooting. Culm cuttings give up to 90% success rates. Small areas of flowering bamboo can be found in most years, and seed is often available.

The combination of multiple uses and ease of propagation by seed or cuttings makes this a highly suitable species for all planting programmes. It also has potential for large scale edible shoot production.

Dendrocalamus hookeri   (Nep. *kalo, bhalu bans*)   D1

Culm &
culm sheath

- brown fur,
  becoming shiny

- round auricle
  with bristles

- dense hairs in
  chevron pattern

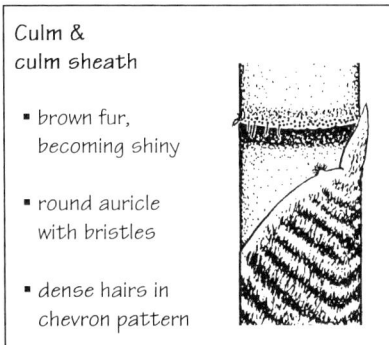

Leaf sheath

- short ligule

- no auricle

- few bristles
  on shoulders

- no hairs

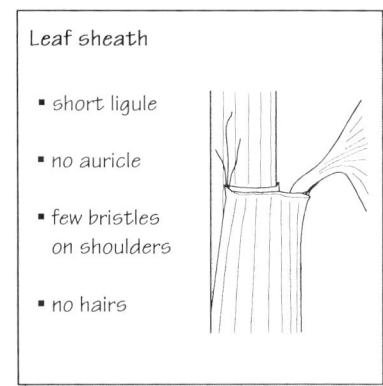

Culm sheath ligule & auricles

- ligule shortly ciliate,
  serrated, 2-4mm wide

- long bristles on auricles

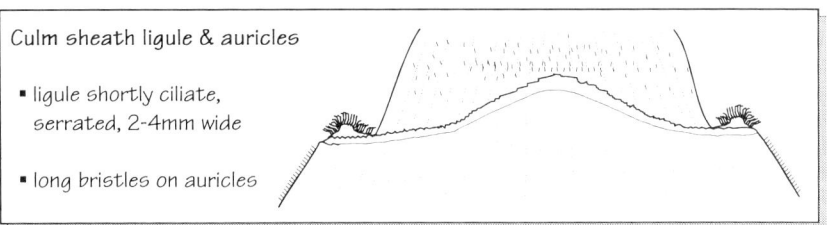

A common, widely-cultivated species of eastern Nepal, similar to *D. sikkimensis*. It can be distinguished from *D. sikkimensis* by the much smaller auricles on the culm sheaths, and by its leaf sheaths, which have fewer bristles. *D. hamiltonii* is similar, but has no bristles on the culm sheath auricles, and longer leaf sheath ligules. *Bambusa clavata* is also similar but has lighter culm sheath hairs and a wider culm sheath ligule. The thick brown hairs on the culm sheaths are often left in a distinctive chevron pattern where they have been rubbed off during growth.

The culms can reach a maximum diameter of 16cm, and a top height of 25m when unthinned, but they are usually 8-9cm in diameter, and 15-18m tall. Culm walls are thin but not very flexible, so the culms are used for general construction, particularly for roofing, rather than weaving. The leaves are large, and can be an important fodder source in winter. Sections of larger culms are used as containers. Although this species appears very similar to *D. asper,* which is widely grown for its edible shoots, the shoots of *D. hookeri* are much more bitter.

Propagation by all the vegetative techniques is easy because of the abundance of aerial roots and very strong branching. This species can even be propagated from the bases of the large branches on their own.

Found from 1,200m -2,000m, mainly in eastern Nepal, rare in central districts.

Dendrocalamus strictus   (Nep. *latthi bans,* Eng. *male bamboo*)   D18/D28

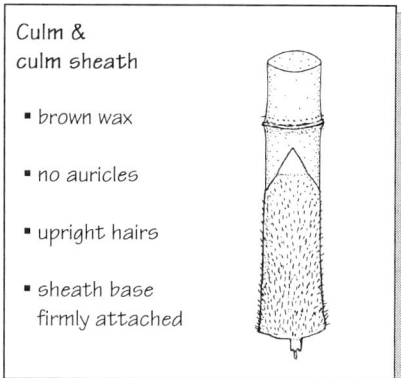

Culm &
culm sheath

• brown wax

• no auricles

• upright hairs

• sheath base
firmly attached

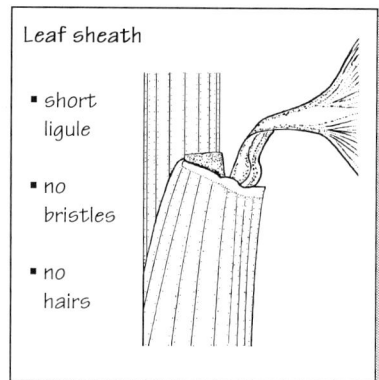

Leaf sheath

• short
ligule

• no
bristles

• no
hairs

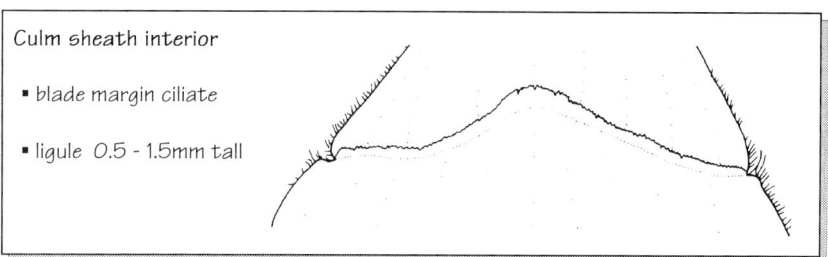

Culm sheath interior

• blade margin ciliate

• ligule 0.5 - 1.5mm tall

The variety native to Nepal is only 6m in height, with a diameter of up to 3cm and completely solid culms. Larger varieties with thinner walls have been planted from Indian seed, and they may appear similar to *Bambusa balcooa.* The thick walls, absence of thorn-like branchlets, and absence of hairs on the leaf sheaths can distinguish this species from *B. balcooa.*

The native variety has few uses. It provides truncheons, often used in the terai for driving animals. The larger varieties provide strong, but short culms, which are widely used for general constructional purposes in India, although the culms are not very straight, and branching is often heavy. Large areas of this species are managed in India to supply paper mills, although the culms are crooked and the pulp is of low quality. Its continued use in India is due to the presence of large natural stands, the availability of seed, and its drought tolerance. It is usually found below 1,000m, although it is often planted at higher altitudes in Nepal.

This species flowers gregariously on a short cycle of 20 to 40 years, and dies after flowering, so it is best to propagate it from seed rather than from cuttings. The seedlings are more drought tolerant than those of other species, and can be raised in beds without shading. A small percentage of seedlings flower in the first few years of growth.

2. BAMBUSA

A genus containing large bamboos of up to 26m in height, as well as several smaller species of only 10m or less. These clump-forming bamboos are similar to *Dendrocalamus* species, but they are generally smaller, with straighter culms and thicker culm walls. The leaves are smaller, and the new culms usually have a thin pale waxy covering rather than dense furry wax. The branches are more uniform in size. Central branches are usually less than 5cm in diameter (fig. 11), so that propagation from culm cuttings can be difficult. Branches are often found right to the base of the culm, and they are thorny in some species. The flowers are in spiky inflorescences (fig. 10 cf. fig. 6), and the bracts at the inflorescence base have two ciliate keels (fig. 12 cf. fig. 8). The species are tropical to subtropical in distribution, occurring up to 1,500m. They have strong thick culm walls which provide a very important source of construction material. They are also sometimes used for weaving.

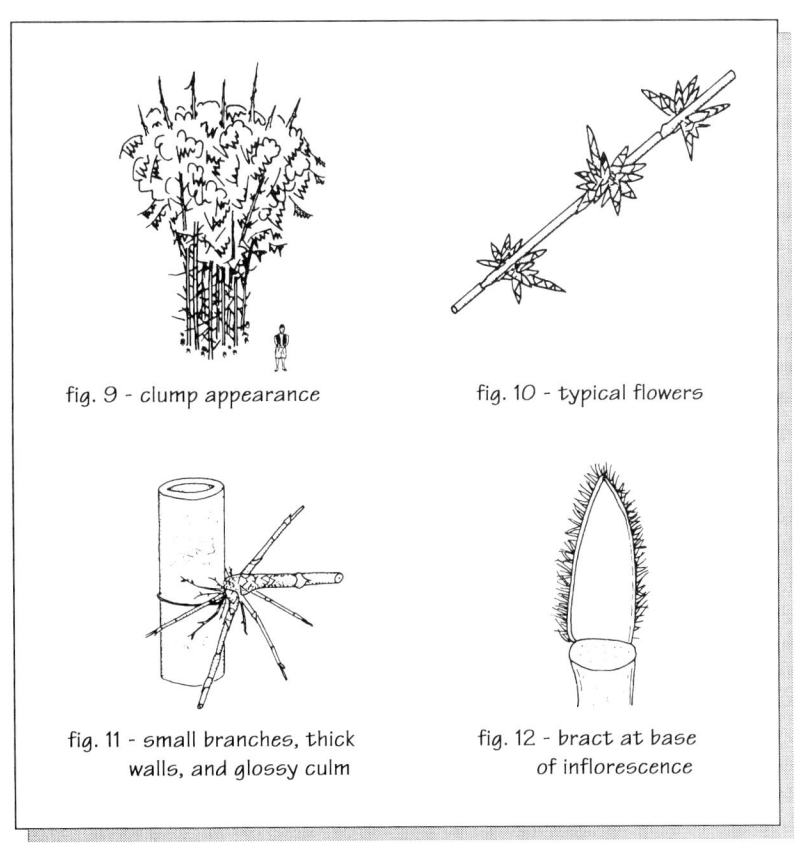

fig. 9 - clump appearance

fig. 10 - typical flowers

fig. 11 - small branches, thick walls, and glossy culm

fig. 12 - bract at base of inflorescence

KEY TO *BAMBUSA* SPECIES

Maximum culm height less than 10m; no hairs on new culm sheaths:-

culm sheath auricles absent or very small, similar .... *multiplex*

culm sheath auricles different, one much larger,
extending down side of sheath .... *alamii*

Maximum culm height more than 10m; new culm sheaths densely hairy:-

Culm sheath auricles large, more than 1.5cm in width or height:-

one culm sheath auricle oval, taller than its width;
bases of some culms with faint yellow stripes; culms
up to 15m tall, slightly crooked, cavity often small and
walls very thick .... *tulda*

both culm sheath auricles wider than their height;
bases of culms without yellow stripes; culms up to
25m tall, very straight, cavity always large:-

culm sheath hairs dark brown; culm sheath blade
cupped and persistent; smaller culms with groove
above branches .... *nutans* subsp. *nutans*

culm sheath hairs jet-black; culm sheath blade very
strongly cupped and deciduous; culms always
round with no groove above branches .... *nutans* subsp. *cupulata*

Culm sheath auricles absent or small, less than 1.5cm in width and height:-

no auricles, even on new culm sheaths; leaf sheaths with
brown hairs .... *balcooa*

culm sheath auricles small, rounded; leaf sheaths with
white hairs .... *nepalensis*

Bambusa alamii   (Nep. *mugi bans*)   B42

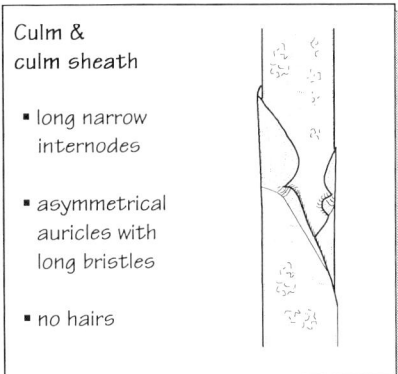

Culm &
culm sheath

- long narrow
  internodes

- asymmetrical
  auricles with
  long bristles

- no hairs

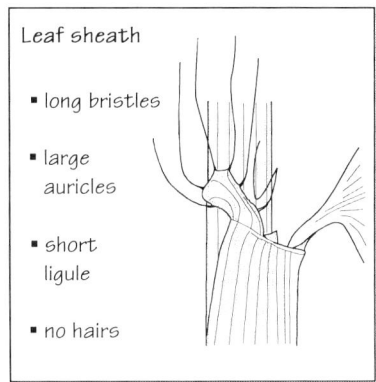

Leaf sheath

- long bristles

- large
  auricles

- short
  ligule

- no hairs

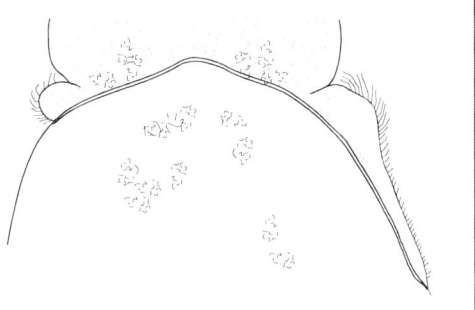

Culm sheath ligule & auricles

- blade separated from auricles

- ligule 1-2 mm wide with
  no serrations on margin

- one long extended auricle,
  one small rounded auricle

This species is similar to Chinese Hedge Bamboo, *Bambusa multiplex*. It is widely cultivated in Bangladesh, and is common in the eastern terai.

Like *B. multiplex* it reaches a maximum diameter of around 4cm and is short for a *Bambusa* species, with a maximum height of 10m. It has very straight culms with long internodes, and little swelling at the nodes.

This species is recognised by its small size, and its glabrous asymmetrical culm sheaths with prominent auricles bearing long bristles. The auricles are separated from the culm sheath blade, and one auricle can extend almost half-way down

the sheath. The leaves have no hairs, and the culm sheath is firmly attached to the culm below the branch bud. The narrow culms with long straight internodes and small branches are highly suitable for splitting into weaving strips.

In the terai small bamboos from genera such as *Drepanostachyum* and *Himalayacalamus* will not grow, and species such as this are a very useful substitute for the smaller hill bamboos, for the production of weaving material.

It is not known whether this species is native to South Asia, or whether it has been introduced from China in the past.

Bambusa balcooa   (Nep. *dhanu bans, ban bans,* Mait. *harod bans*)   D23

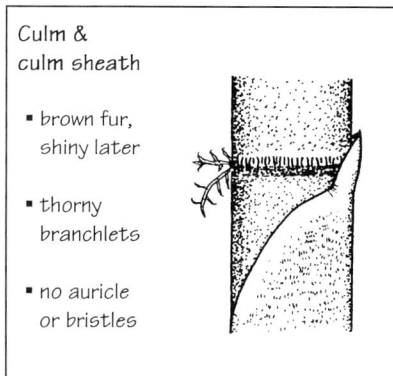

Culm &
culm sheath

- brown fur,
  shiny later

- thorny
  branchlets

- no auricle
  or bristles

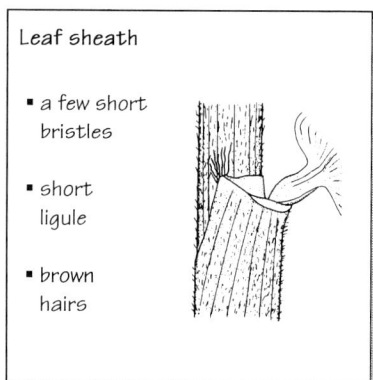

Leaf sheath

- a few short
  bristles

- short
  ligule

- brown
  hairs

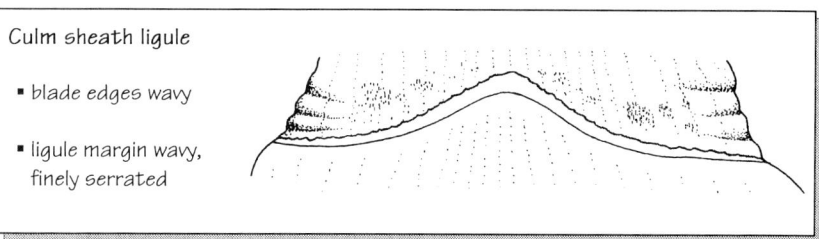

Culm sheath ligule

- blade edges wavy

- ligule margin wavy,
  finely serrated

This is a large thick-walled bamboo with strong branching, and thorn-like branchlets lower down the culm. It can reach a diameter of 16cm and a height of up to 25m. It is similar to species of *Dendrocalamus,* having thick furry culm wax, densely hairy culm sheaths and large branches. It is easy to recognise because of the brown hairs on the leaf sheaths and the small curving thorn-like branchlets. The thorns are smaller than those of *B. bambos,* and there are fewer hairs inside the culm sheath blade. It can be separated from all other large bamboos in the region by the absence of auricles on the culm sheaths.

The poles are highly valued in India, where they are an important raw material which can be marketed in large quantities. They are used for scaffolding and for weaving into panels for making house walls. They are generally a little too large for village use, and the heavy branching makes them difficult to split by hand. They are reserved for a few village uses such as pillars and beams.

This is an adaptable species, widespread across West Bengal and Assam, and it grows well from Calcutta up to around 1,600m. It tolerates drier conditions better than many bamboos, but can suffer from the bamboo blight syndrome on poorer sites.

The large size of this species along with its thorny branchlets make it a good choice for slope stabilisation. The large branches make it easy to propagate from culm cuttings.

Bambusa multiplex (Eng. Chinese hedge bamboo) B23

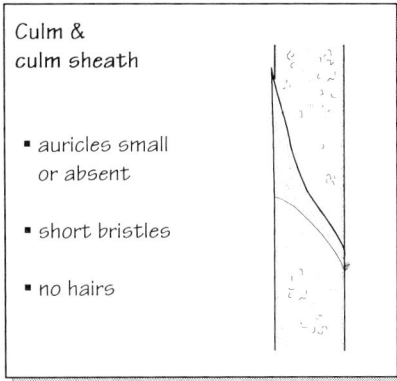

Culm &
culm sheath

- auricles small
  or absent

- short bristles

- no hairs

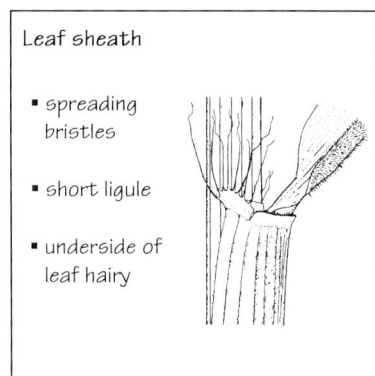

Leaf sheath

- spreading
  bristles

- short ligule

- underside of
  leaf hairy

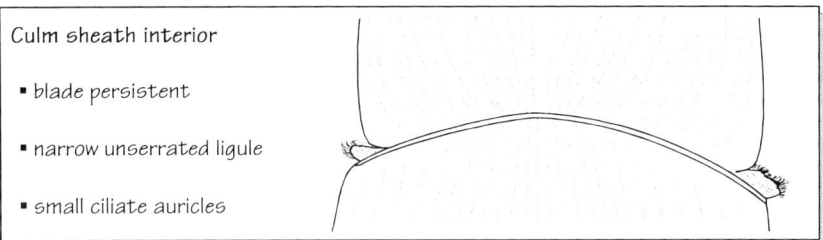

Culm sheath interior

- blade persistent

- narrow unserrated ligule

- small ciliate auricles

This small species, introduced from China, is similar to *Bambusa alamii*. It is cultivated throughout the tropics, with many different ornamental varieties. The common variety in Nepal is up to 10m in height with culms of up to 4cm in diameter. This is the plant known as *B. multiplex* cultivar ' Fernleaf '. Other varieties in Nepal include a miniature plant with tiny leaves reaching only 2m in height, known as the Chinese Goddess bamboo, var. *rivierorum*.

It is easy to confuse these bamboos with species from small subtropical and temperate genera. The branching is superficially very similar to that of species of *Himalayacalamus*, but the broad culm sheath blades distinguish these species quite clearly. This species can be separated from other *Bambusa* species by its narrow culms and small culm sheath auricles. Larger plants have densely hairy undersides to the leaves.

This species, like *B. alamii*, is well suited to cultivation in the terai for production of small weaving material. It is adaptable, growing from Calcutta up to altitudes of 2,000m, and does not appear to flower gregariously. Because of its small branches, culm cuttings would not be successful for this species. The traditional propagation technique is quite easy as the rhizomes are small. In addition branches often develop into rhizomatous offsets with long roots. These can be used for propagation if they are removed and transplanted into containers during the monsoon.

Bambusa nepalensis (Nep. *tama bans, phusre bans*)   D13

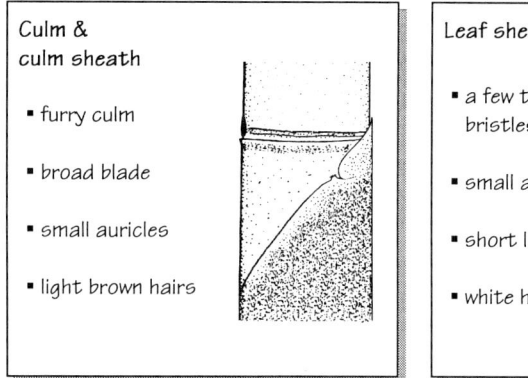

Culm &
culm sheath

- furry culm

- broad blade

- small auricles

- light brown hairs

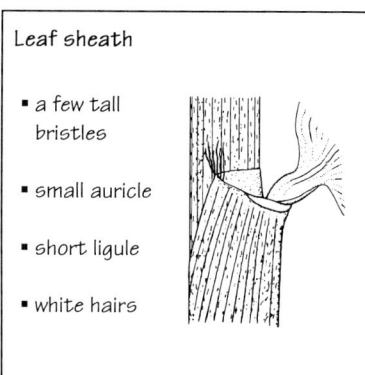

Leaf sheath

- a few tall
  bristles

- small auricle

- short ligule

- white hairs

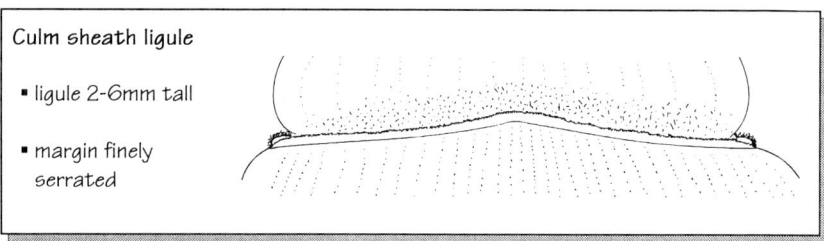

Culm sheath ligule

- ligule 2-6mm tall

- margin finely
  serrated

A widely cultivated species, common from East Nepal to Tansen in the west. the culms reach a maximum diameter of 10cm and a maximum height of 20m. It is similar to *Dendrocalamus hamiltonii*, and is known by the same local name, tama bans, in the Kathmandu Valley. It is a multipurpose species, used for weaving, general construction, and it also produces edible shoots.

The large leaves and dull culms distinguish it from *Bambusa nutans*. The culm sheath hairs are lighter than those of *Dendrocalamus hamiltonii* and *D. hookeri*. Unlike *Dendrocalamus giganteus* and *Bambusa balcooa* it has small ciliate culm sheath auricles, and white leaf sheath hairs. The culm sheaths are very broad and the blade is flattened tightly against the culm. The hairs on the culm sheath are short, dense, and flattened, giving a smooth furry appearance.

There are few branches in the lower half of the culm, and the nodes are not raised. This makes the culms easy to split and gives straight sections for weaving.

Mid-culm central branches are large, often with aerial roots, and propagation is easy by both the traditional technique, and by culm cuttings. However, the absence of branches lower down the culm makes it essential to use a long pole in the traditional cutting, and also reduces the number of culm cuttings which can be made from each culm. Sporadic flowering of isolated clumps is common, and seedlings have been raised in several forest nurseries.

Bambusa nutans subsp. **nutans** (Nep. *tharu bans, sate bans*)   B21

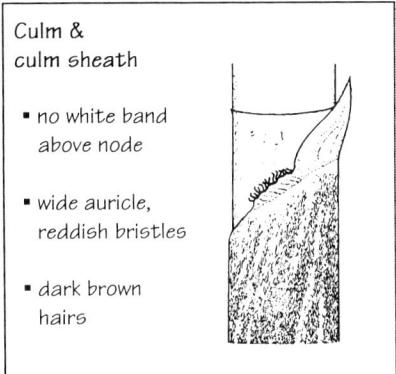

Culm &
culm sheath

• no white band
  above node

• wide auricle,
  reddish bristles

• dark brown
  hairs

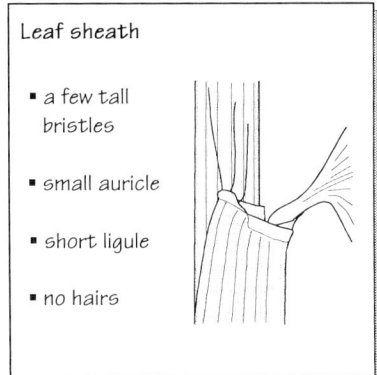

Leaf sheath

• a few tall
  bristles

• small auricle

• short ligule

• no hairs

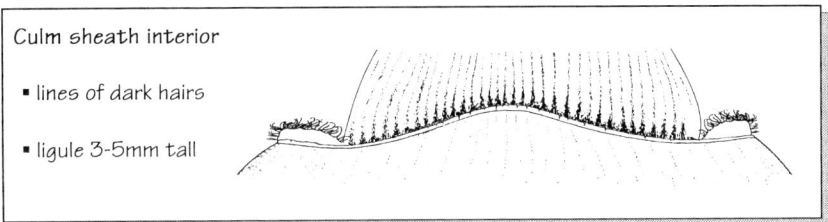

Culm sheath interior

• lines of dark hairs

• ligule 3-5mm tall

The commonest cultivated bamboo in the hills of central and west Nepal. It is not found in the terai. In its large culm sheath auricles it is similar to both *B. nutans* subsp. *cupulata,* and to *B. tulda.* Leaf sheath auricles are smaller than those of *B. tulda.* This subspecies can be separated by its weakly cupped and persistent culm sheath blades, and brown culm sheath hairs. Small culms are often flattened on one side above each bud or branch cluster, while those of subsp. *cupulata* are completely round.

The poles reach a maximum diameter of 10cm and are up to 23m long. They are strong and highly prized for all constructional purposes, and are reputed to be resistant to termite attack. They can also be used for the weaving of rough baskets and mats as the branches are small and the poles split easily. The poles are used for carrying corpses to the funeral pyre. The shoots are bitter and are not eaten. It will tolerate dry sites well, and can lose most of its leaves in the spring drought without harm.

This subspecies roots well from culm cuttings if strong culms are chosen and planted at the right time, even though the branches are quite small and there are few aerial roots. Success rates of 75% have been achieved. It is also easy to establish this subspecies by the traditional technique.

Sporadic flowering of individual clumps is common, but seed has never been found. The spikelets are often filled with an orange or black fungus.

Bambusa nutans subsp. **cupulata** (Nep. *mal bans*) B1

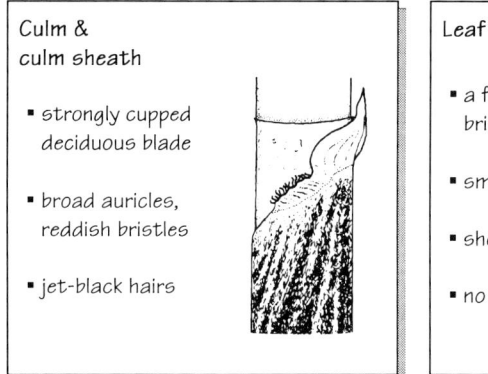

Culm &
culm sheath

• strongly cupped
deciduous blade

• broad auricles,
reddish bristles

• jet-black hairs

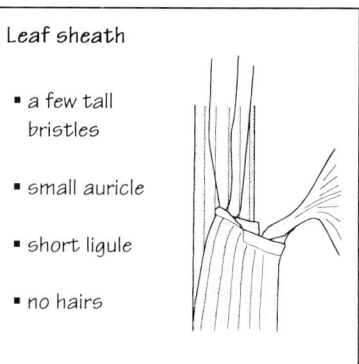

Leaf sheath

• a few tall
bristles

• small auricle

• short ligule

• no hairs

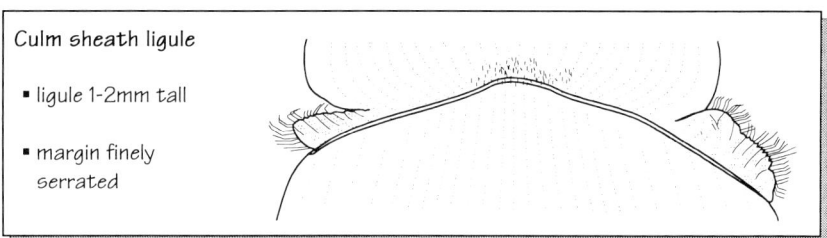

Culm sheath ligule

• ligule 1-2mm tall

• margin finely
serrated

The commonest cultivated bamboo east of Okhaldhunga and Malangwa, from the terai up to 1,500m. It is similar to *B. tulda* and *B. nutans* subsp. *nutans* in its large culm sheath auricles. This subspecies can be recognised by its strongly cupped deciduous culm sheath blades, and jet-black culm sheath hairs.

The poles reach a maximum diameter of 10cm and are up to 23m long. They are strong and highly prized for all constructional purposes. They can also be used for the weaving of rough baskets and mats as the branches are small and the poles split easily. The culms are very straight and the unraised nodes with small branches and small leaves make it a very attractive and clean-looking bamboo, especially in the hills, where a

taller form with smaller branches is cultivated. The leaves are widely used for fodder. The shoots are bitter and are not eaten. This is one of the most desirable bamboos for many end-uses, having long straight culms and small branches. It will tolerate dry sites well, and it is common from the eastern Himalayas to the Bay of Bengal.

This subspecies does not generally root well from culm cuttings as the branches are small and there are few aerial roots. However, at lower altitudes there are forms with shorter culms and larger branches which can be more successful. It is easiest to establish this subspecies by the traditional technique. Rooting branch clusters have not yet been successfully used for propagation.

Bambusa tulda   (Nep. *kada bans, koraincho bans, chab bans*)   B22

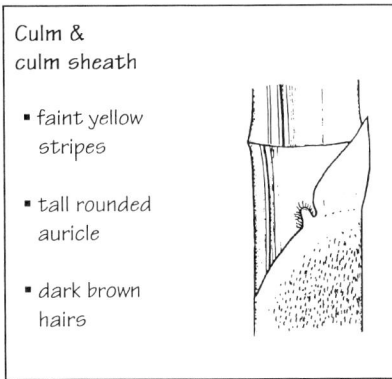

Culm &
culm sheath

- faint yellow
  stripes

- tall rounded
  auricle

- dark brown
  hairs

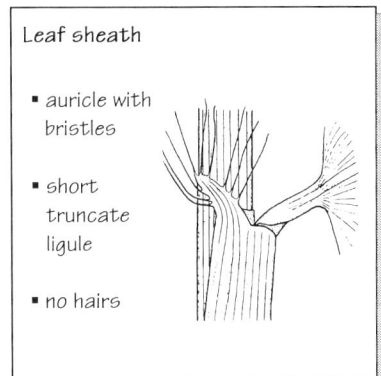

Leaf sheath

- auricle with
  bristles

- short
  truncate
  ligule

- no hairs

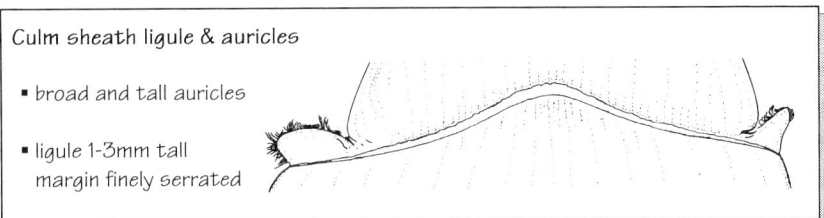

Culm sheath ligule & auricles

- broad and tall auricles

- ligule 1-3mm tall
  margin finely serrated

This species is rare in the Himalayas. It is occasionally found in the terai, especially around the Chitwan area. It is also found along paths leading from the terai into the hills, and in the Kathmandu valley. It has strong upright culms, but they are quite short and can be rather crooked, with swollen nodes and very heavy branching.

B. tulda can be distinguished from *B. nutans* by the larger, more prominent, leaf sheath auricles. There are often faint yellow stripes on the lower internodes of some culms, and one culm sheath auricle is often quite tall. The hairs on the culm sheath are not as black as those of *mal bans*, and the culm sheath blade is less cupped, and more persistent. It is similar to *B. vulgaris*, which also has varieties with striped culms, but it has no bristles on the lower edges of the culm sheath blade and larger leaf sheath auricles. The culms are more crooked than those of other *Bambusa* species with much thicker walls.

The culms can reach a maximum diameter of 7cm and a length of 15m, although they are often smaller. As they are very thick-walled they are used for constructional purposes. Leaves can be used for fodder, but they are small. The shoots are not edible. The thick walls and strong branching make it easy to propagate this species by any vegetative means, and branches on their own may root successfully, but either subspecies of *Bambusa nutans* will give longer straighter culms with lighter branching.

3. THAMNOCALAMUS

Clump-forming thornless frost-hardy bamboos, up to 5m tall, found from 2,800m to 3,500m in temperate forest, with cross-veined leaves, smooth or waxy culms, few branches, and usually with upright culm sheath blades. These are the highest altitude clump-forming bamboos, growing above the range of the other forest clump-forming genera *Cephalostachyum*, *Himalayacalamus*, and *Drepanostachyum*. They can always be distinguished from those genera by the prominent cross-veins on their leaves

(fig. 15). In distinction to *Borinda* the branches are fewer, and do not extend behind the culm in the first year (figs 14 and 20), and the culms are smooth and straight. Buds on the culms are tall (fig. 16), unlike the short buds found in *Himalayacalamus* and *Drepanostachyum* species. The rhizomes are solid, shorter than those of *Yushania* and *Arundinaria*, less than 30cm long (fig. 15). The culms are small and brittle and not widely used, but the shoots and leaves provide important food and cover for wildlife.

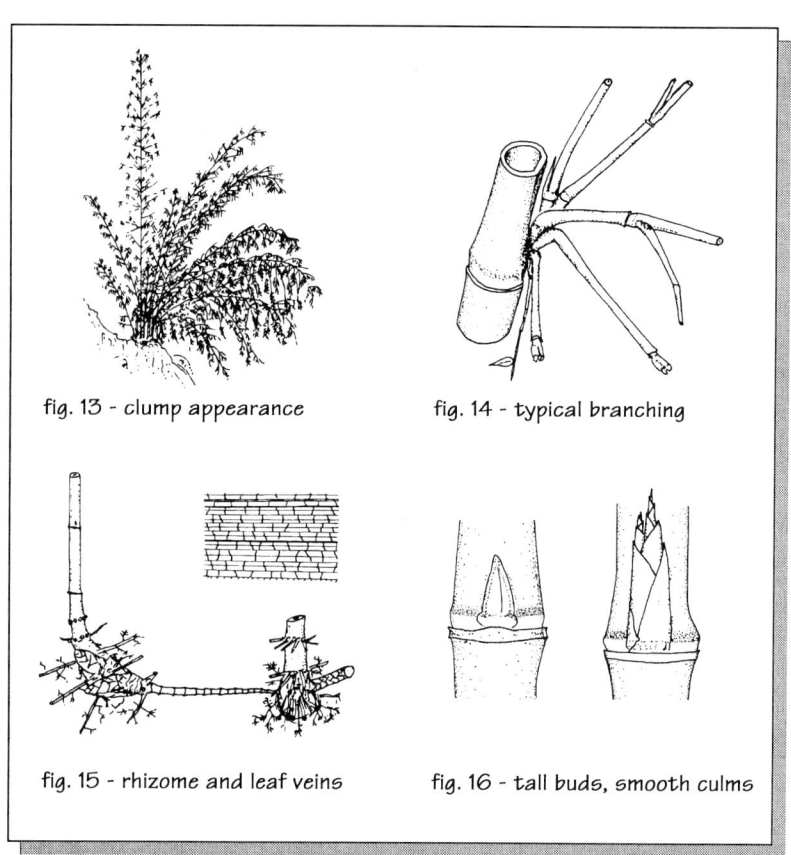

fig. 13 - clump appearance

fig. 14 - typical branching

fig. 15 - rhizome and leaf veins

fig. 16 - tall buds, smooth culms

KEY TO *THAMNOCALAMUS* SUBSPECIES AND VARIETIES

Culm sheath shoulders broad with many bristles; culm nodes
strongly swollen; branching heavy (Langtang, Helambu) .... var. *crassinodus*

Culm sheath shoulders narrow with few bristles; culm nodes
not strongly swollen; branching light:-

    new culm sheaths densely hairy; leaves narrow
    (central to east Nepal) .... subsp. *spathiflorus*

    new culm sheaths with no hairs; leaves broad,
    (central to west Nepal) .... subsp. *nepalensis*

Thamnocalamus spathiflorus subsp. **spathiflorus** (Nep. *rato nigalo*) T31/T30

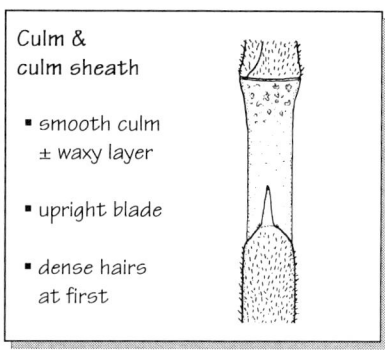

Culm &
culm sheath

- smooth culm
  ± waxy layer

- upright blade

- dense hairs
  at first

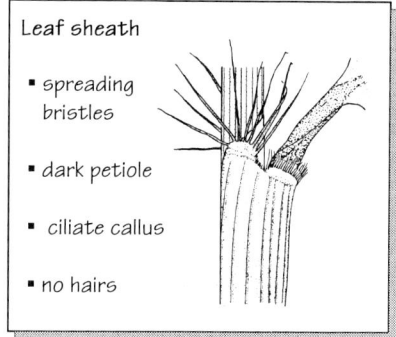

Leaf sheath

- spreading
  bristles

- dark petiole

- ciliate callus

- no hairs

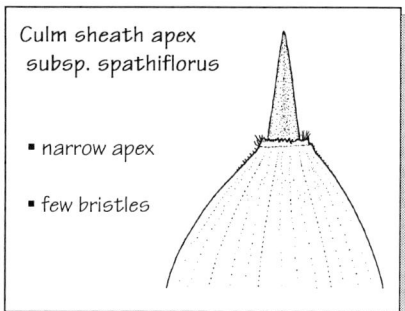

Culm sheath apex
subsp. spathiflorus

- narrow apex

- few bristles

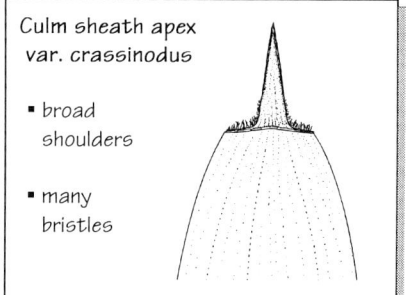

Culm sheath apex
var. crassinodus

- broad
  shoulders

- many
  bristles

A common bamboo of central and eastern Nepal, found between 2,800m and 3,500m, usually above *Yushania maling* or species of *Himalayacalamus* or *Borinda*. It prefers steeply sloping sites. This species extends right along the Himalayas and has several subspecies and varieties. The leaves have prominent cross-veins, and are often on long pendulous branchlets with many short internodes.

This subspecies is distinguished from the western subspecies *nepalensis* by its hairy culm sheaths and the bristles on the leaf sheaths. The leaves are narrow, usually in groups of 5 to 9 on each branchlet. Old culms become red. In Langtang and Helambu areas

the variety *crassinodus* is found. It has swollen culm nodes and small leaves. The culm sheaths are hairy or glabrous with broad ciliate shoulders.

This species is not harvested if larger bamboos such as *Himalayacalamus* species are available. The small brittle culms with swollen nodes make it unsuitable for weaving. However, in some areas it is the only forest bamboo, and it is harvested annually. It is extremely important for wildlife, providing food for animals such as red pandas and bears, and shelter for birds such as pheasants. It is also browsed by livestock in winter. It does not hinder regeneration of trees, as seedlings can grow in the gaps between the clumps.

Thamnocalamus spathiflorus subsp. **nepalensis** (Nep. *jarbuto*) T26/T30

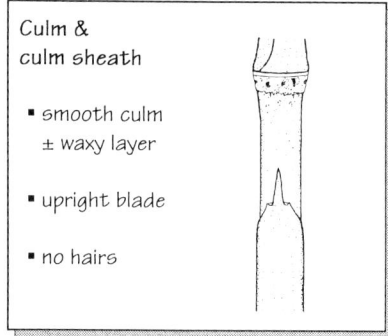

Culm &
culm sheath

- smooth culm
  ± waxy layer

- upright blade

- no hairs

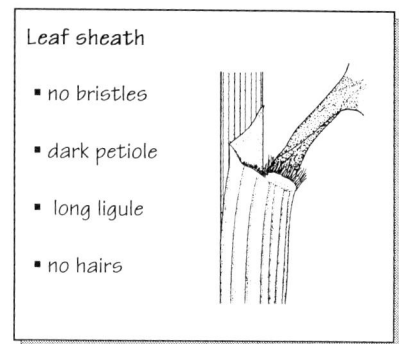

Leaf sheath

- no bristles

- dark petiole

- long ligule

- no hairs

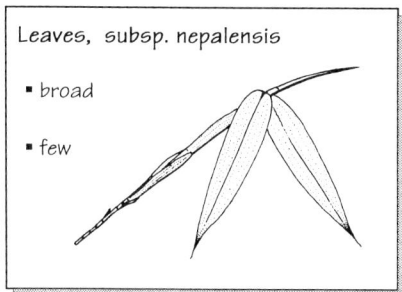

Leaves, subsp. nepalensis

- broad

- few

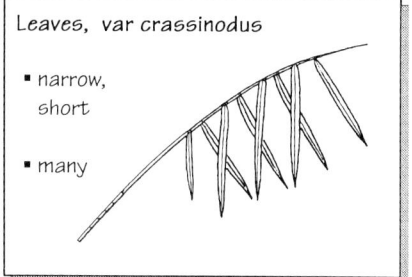

Leaves, var crassinodus

- narrow,
  short

- many

A common bamboo of central and western Nepal found between 2,800m and 3,500m, usually above species of *Himalayacalamus* or *Borinda*. It prefers steeply sloping sites. This species extends right along the Himalayas and has several subspecies and varieties. Old culms have very long branchlets with many short internodes, and the leaves have prominent cross-veins.

This western subspecies, *nepalensis,* is distinguished from *spathiflorus,* the eastern subspecies, by its hairless culm sheaths and the lack of bristles on the leaf sheaths. The leaves are broad, and are usually in groups of 2 to 5 on each branchlet. The culms turn purple or black with age, especially when growing in exposed sites. In Langtang and Gosainkund areas the variety

crassinodus is common. It has swollen culm nodes and smaller leaves. The culm sheaths may be hairy or glabrous and have broad ciliate shoulders.

This species is not harvested if larger bamboos such as *Himalayacalamus* species are available nearby. It has small culms which are brittle and have swollen nodes, which make it unsuitable for weaving. In some areas it is the only forest bamboo, and is harvested annually. It is extremely important for wildlife, providing food for animals such as red pandas and bears, and shelter for birds such as pheasants. It is also browsed by livestock in winter. It does not hinder regeneration of trees, as seedlings can grow in the gaps between the clumps.

4. BORINDA

Clump-forming frost-hardy bamboos, found in temperate forest from 1,800-3,200m, up to 10m tall, with prominent cross-veins on the leaves, tall buds, and culms which are either finely grooved or curve outwards at the base. These bamboos are quite similar to *Thamnocalamus*, but grow at a lower altitude, and have more branches. There are up to seven branches in the first year, two branches extending behind the new culm (fig. 16). The grooved or curving culms separate *Borinda* from the genera *Thamnocalamus*, *Himalayacalamus*, and *Drepanostachyum*, which have smooth upright culms. The culms are similar in their finely-grooved surface to those of *Cephalostachyum* and *Ampelocalamus*, but the buds are taller and the leaves have cross-veins. The rhizomes are less than 30cm long, shorter than those of the spreading genera *Yushania* and *Arundinaria*, (fig. 15). These rare bamboos are not generally harvested. Tree seedlings can easily regenerate between the clumps.

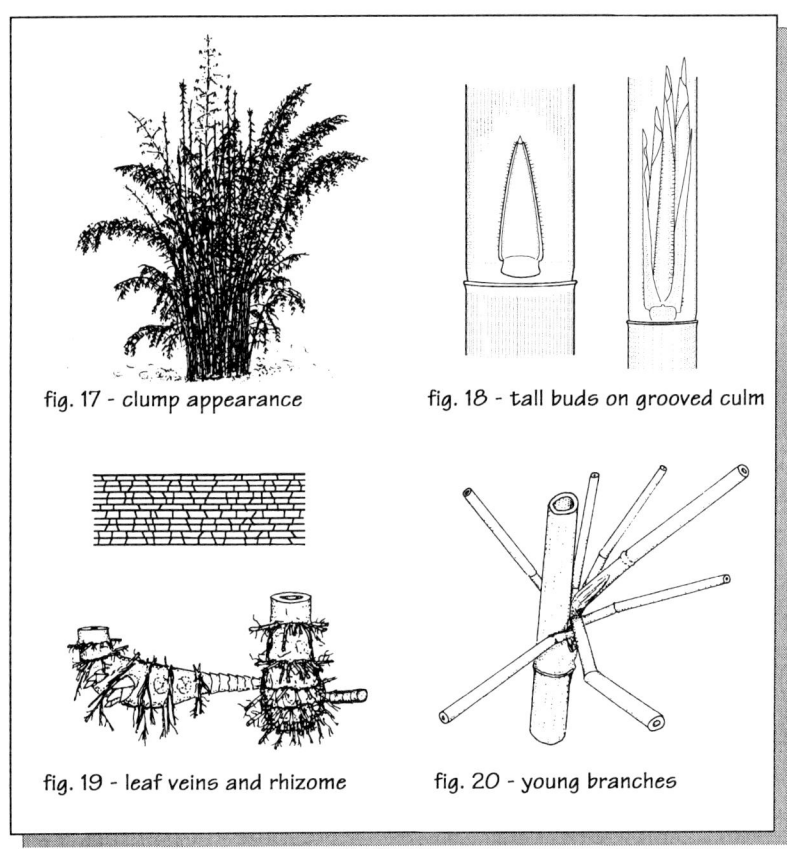

fig. 17 - clump appearance

fig. 18 - tall buds on grooved culm

fig. 19 - leaf veins and rhizome

fig. 20 - young branches

KEY TO *BORINDA* SPECIES IN NEPAL

Culms curving outwards and upwards at base; surface smooth without
fine grooves; culm sheaths fragile without clearly defined blade
and with very tall narrow ligules; leaves narrow .... *chigar*

Culms straight and upright at base; culm surface with fine grooves;
culm sheaths tough with clearly defined deciduous blade and
broad ligule; leaves broad .... *emeryi*

Borinda chigar   (Nep. *chigar*)   T25

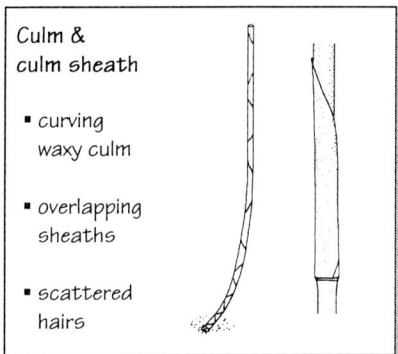

Culm &
culm sheath

- curving
  waxy culm

- overlapping
  sheaths

- scattered
  hairs

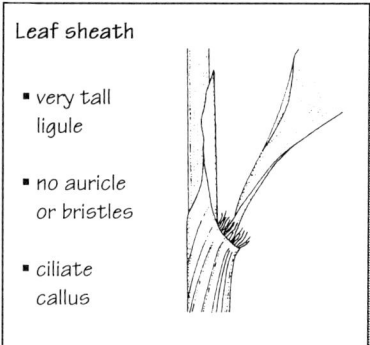

Leaf sheath

- very tall
  ligule

- no auricle
  or bristles

- ciliate
  callus

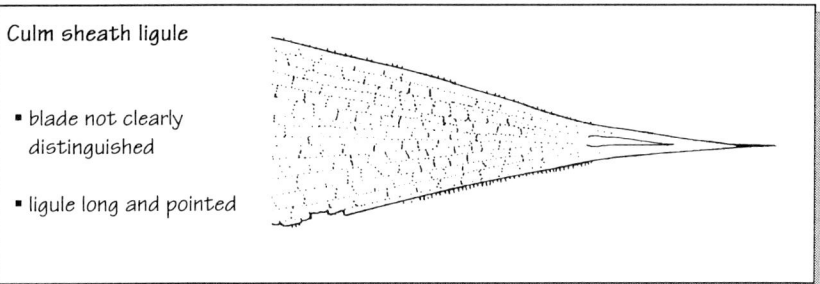

Culm sheath ligule

- blade not clearly
  distinguished

- ligule long and pointed

A locally common species, found between *Himalayacalamus cupreus* and *Thamnocalamus spathiflorus* from 2,600m to 3,100m. It has only been found in one location so far, near Machhapuchare in Kaski district of West Nepal, but it may be more widespread.

This species differs from all other Nepalese species so far encountered in its curving culm bases, which lead to greater separation of the culms, and a very open form of clump. It also has distinctive culm sheaths and leaves. The culm sheaths overlap, and are very fragile. They have long tapering ligules inside blades which are not well distinguished from the rest of the sheath. The leaves are very narrow, and the leaf sheaths also have very long ligules. Branching is light with long slender branchlets, and possibly fewer branches at each node than other *Borinda* species. The culms are covered in a light white wax, which is persistent, the culms remaining dull and not shiny.

This bamboo is not used for weaving because of the curved culms. However, because the clumps are quite open, it provides valuable protection and shelter for wildlife such as pheasants.

More information is required on this species. It has been placed in *Borinda* but this is rather speculative as its flowers are not yet known. It appears to have similar culms and culm sheaths to other species from Bhutan, and Yunnan Province of South-west China.

Borinda emeryi   (Nep. *kalo nigalo*)   T59

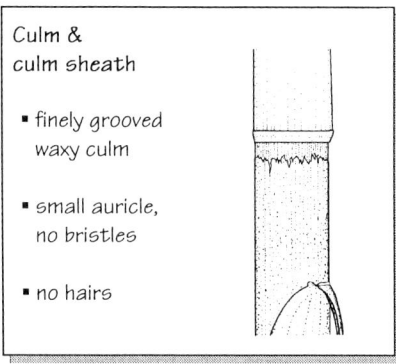

Culm &
culm sheath

- finely grooved
  waxy culm

- small auricle,
  no bristles

- no hairs

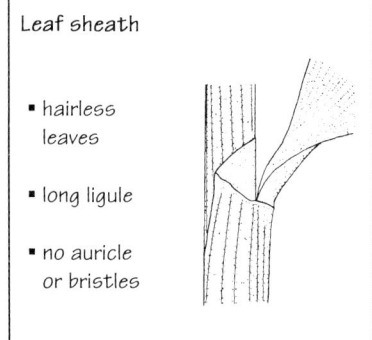

Leaf sheath

- hairless
  leaves

- long ligule

- no auricle
  or bristles

Culm sheath apex

- small hairy auricles, no bristles

- ligule wide, short, and ciliate,

- persistent blade with few hairs

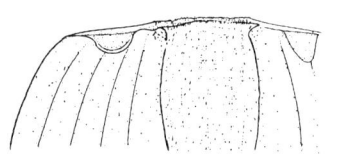

A rare species of East Nepal, collected mainly in the Barun Khola valley of Sankhuwasabha district, but also found along the Milke Dande, between 2,600m and 3,200m. It is naturally restricted to wetter temperate mixed coniferous and rhododendron forest areas of East Nepal, and is apparently not cultivated.

This species has been collected on several occasions, but never with full-sized culms or with new culm sheaths, so it is difficult to describe it precisely. It is closely related to better known species from Bhutan and Tibet, which reach large dimensions for the altitude at which they grow, and are extremely important minor forest products. They have culms up to 10m tall and up to 4.5cm in diameter, and

leaves up to 25cm long. This species is easily distinguished from the other frost-hardy bamboos by its finely grooved culms which have a persistent waxy covering which turns black with age. The culm sheaths have no hairs. *Thamnocalamus spathiflorus* is similar, but has smooth reddish culms and hairy sheaths. *Yushania maling* is also similar, but has rough culms, long rhizomes and hairy sheaths with bristles.

Because of their level nodes, thin walls, and long internodes, the culms would split easily into strips for weaving. Cultivation of this species is only possible by the traditional technique. Very long poles should be used for propagation as there are few branch buds at the base of the culm.

5. AMPELOCALAMUS

Clump-forming thornless bamboos, up to 12m tall, cultivated from 1,200m to 2,000m, with long internodes, no cross-veins on the leaves, and short buds. This genus is very similar in appearance to two other medium stature subtropical clump-forming genera, *Cephalostachyum* and *Teinostachyum*, although their branches and inflorescences show that they are not closely related at all. *Ampelocalamus* species have finely ridged culm internodes of more than 40cm, short broad buds, no cross-veins on their leaves, and long pendulous culm tips. They can be distinguished from *Cephalostachyum* species in Nepal by their distinctive branching. The branches are all similar in size, and are arranged in vertical groups. They curve outwards from the culm and have swollen bases, and the larger branches often bear aerial roots. Most species are found only in China, but one species extends from Yunnan along the Himalayas to central Nepal. It is probably introduced, as it has not yet been found in forest areas.

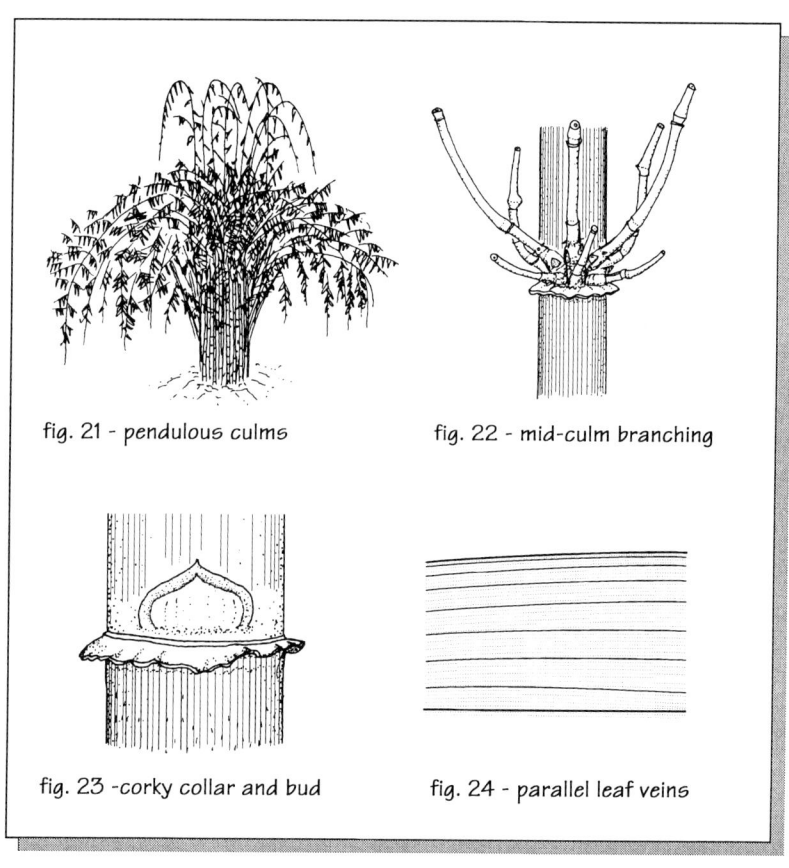

fig. 21 - pendulous culms

fig. 22 - mid-culm branching

fig. 23 -corky collar and bud

fig. 24 - parallel leaf veins

Ampelocalamus patellaris   (Nep. *nibha, ghopi bans, lyas bans*)   T3

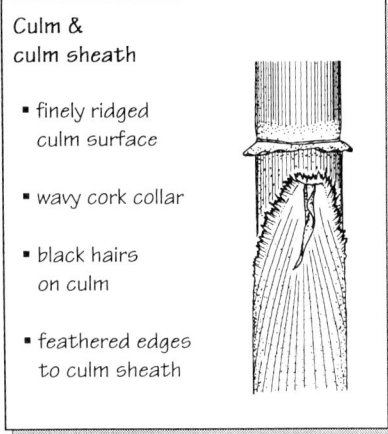

Culm &
culm sheath

• finely ridged
  culm surface

• wavy cork collar

• black hairs
  on culm

• feathered edges
  to culm sheath

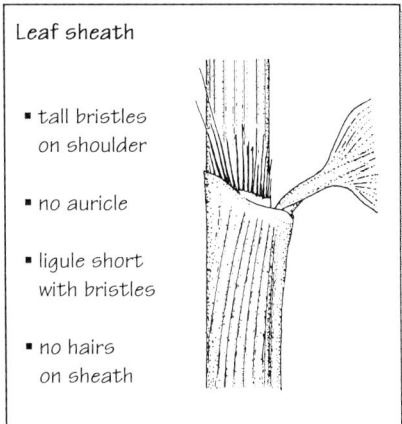

Leaf sheath

• tall bristles
  on shoulder

• no auricle

• ligule short
  with bristles

• no hairs
  on sheath

A useful and also very attractive cultivated species, with poles reaching 5cm in diameter and 12m in height. In its natural forest environment this is a scrambling bamboo, but it is usually cultivated in self-supporting clumps.

This is a very easy species to recognise, as the culm sheaths have distinctive long-fringed edges at the top. The leaf sheaths have no auricles, but they have a few very upright bristles and the edge of the ligule also has long bristles and cilia. The culms also make this species easy to recognise as they have a distinctive corky collar around each node. This helps to support the flexible upper sections of the culms as they straggle over tree branches.

As it is such a pendulous bamboo the culms are very flexible and the long internodes of up to 50cm make them very useful for weaving. This is the main use of this bamboo, as the culm walls are too thin for the culms to be of structural value. The leaves are large, up to 40cm long, and can be used as fodder.

The branches are irregular in shape with curving internodes and swollen nodes. This allows re-orientation towards the light, and helps to support the scrambling branches. The central branch is only slightly larger than the rest and it often bears aerial roots. Propagation by culm cuttings should be feasible because of these aerial roots.

This species was first described with the name *Dendrocalamus patellaris*, but it is now known that the flowers which were originally collected came from a clump of *Dendrocalamus hamiltonii* instead. It occurs from 1,200 to 1,800m, especially in higher rainfall areas, particularly Ilam and Taplejung disticts of East Nepal, and Kaski and Palpa sistricts of West Nepal. A widespread flowering ocurred around 1980 in East Nepal and Darjeeling district of West Bengal.

6. CEPHALOSTACHYUM

Clump-forming tropical and subtropical bamboos up to 10m tall and up to 5cm in diameter, found in high rainfall forests. Smaller than *Bambusa* and *Dendrocalamus* species but larger than temperate bamboos, this genus is useful for weaving into mats, having flexible culms with internodes up to 1m long. The long internodes, often with fine ridges, are similar to those of *Borinda emeryi*, but the leaves only have long parallel veins (fig. 28) without any of the short cross-veins seen in frost-hardy bamboos, (fig. 15). The buds are short and rounded, (fig. 27), while those of the frost-hardy genera are tall and narrow, (fig. 16). They are also similar to *Ampelocalamus patellaris*, but do not have the frilly collar at the nodes or the long-fringed culm sheath margins. The tips of the culms are long and thin and hang down to the ground or sprawl over tree branches. Culm sections can be made into flutes. They are usually found in natural forest and have not been seen in cultivation in Nepal.

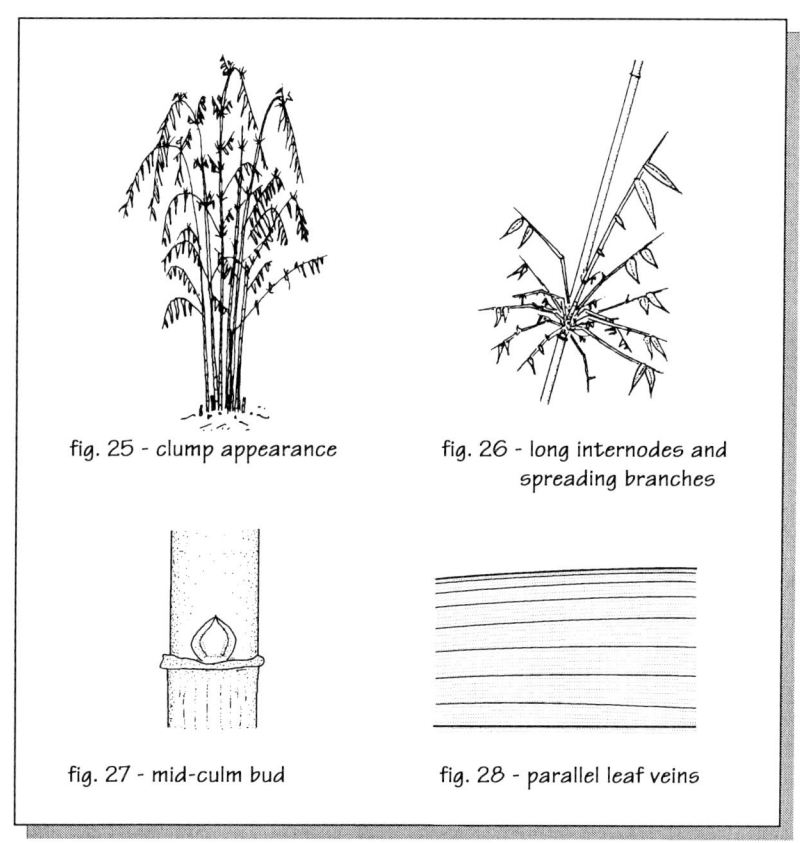

fig. 25 - clump appearance

fig. 26 - long internodes and spreading branches

fig. 27 - mid-culm bud

fig. 28 - parallel leaf veins

Cephalostachyum latifolium   (Nep. *ghopi bans, murali bans*)   U43

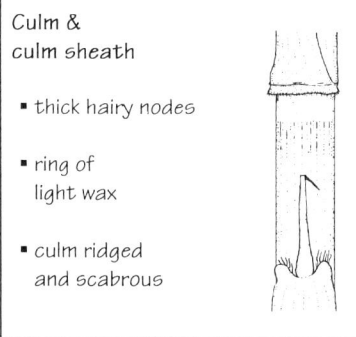

Culm &
culm sheath

- thick hairy nodes

- ring of
  light wax

- culm ridged
  and scabrous

Leaf sheath

- tall flattened
  white bristles

- ligule long
  and glabrous

- sheath edge thin
  with cross-veins

Culm sheath apex

- tall flattened white bristles

- raised delicate shoulders
  with visible cross-veins

- ligule very short, blade
  with dense brown hairs

A distinctive but rare species of the cooler subtropical forests of central and eastern Nepal, usually found from 1,500 -2,000m. Culms are up to 15m long and 5cm in diameter, with internodes of up to 1m. The straggling clumps have long pendulous culm tips, and very large leaves for the size of the culms.

The ridged culm sheaths with thin edges and tall shoulders distinguish this from a similar species in West Bengal and Bhutan, *Teinostachyum dullooa*. The leaf sheaths also have tall shoulders, and both culm and leaf sheaths have long white bristles when young. The bristles are delicate and deciduous, leaving hardly any trace once they have fallen.

Culm nodes and sheath bases have short light brown hairs. The culm nodes are swollen, with a corky collar similar to that of *Ampelocalamus patellaris,* but much thinner

The flexible culms with long internodes are very useful for weaving and making flutes, and the broad leaves, up to 30cm long and 6cm wide, make excellent animal fodder. This bamboo is often harvested from the forest on a regular basis. Most of this species flowered in Bhutan in recent years, and many seedlings and small young clumps are now encountered in the forest. The seed are large, and the flowering branches were used as paint brushes.

7. DREPANOSTACHYUM

Clump-forming thornless bamboos up to 5m tall with many branches, found from 1,000-2,200m in dry subtropical forests, and also cultivated. Leaves have no cross-veins (fig. 30), and culm internodes are less than 40cm long. Branch buds at the nodes of the culms are shorter than their height, and always open (fig. 31). The buds have many small initials visible, which will produce up to 70 branches at each node, about 25 growing in the first year. The branches are quite uniform in size and spread around the culm (fig. 32). When growing vigorously the upper half of the culm sheaths are very narrow and the culm sheath ligules are long and ragged. The sheaths are always rough inside at the top, and this distinguishes them clearly from species of *Himalayacalamus*. Rhizomes are short and solid, less than 30cm in length, and similar to those of *Thamnocalamus* and *Borinda*. The culms are valuable for weaving and the foliage is often fed to animals or browsed in the forest. The new shoots are very bitter.

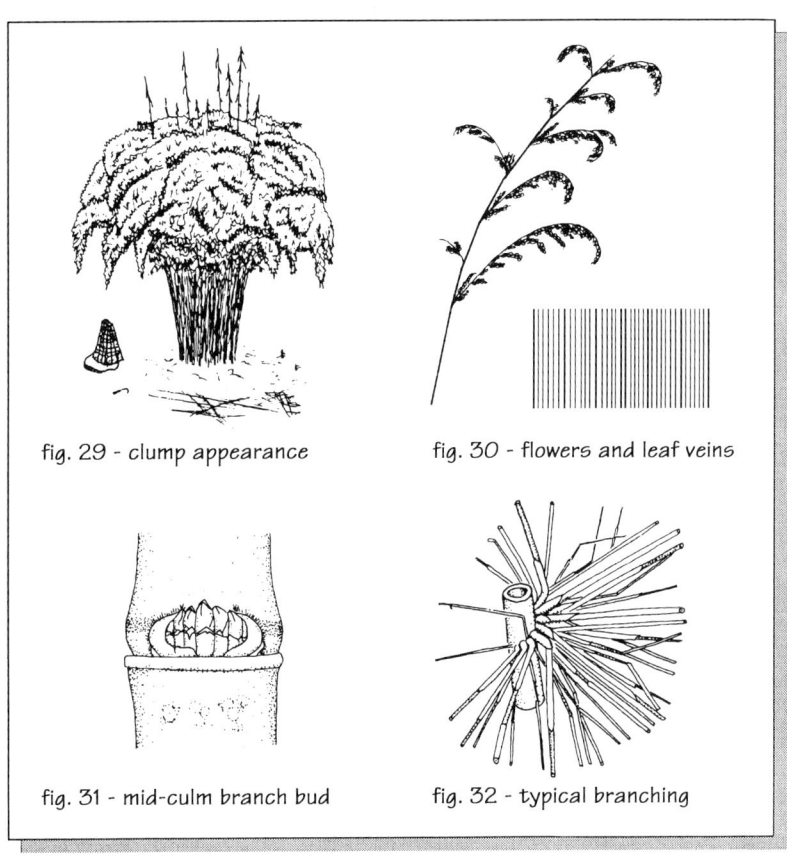

fig. 29 - clump appearance

fig. 30 - flowers and leaf veins

fig. 31 - mid-culm branch bud

fig. 32 - typical branching

KEY TO *DREPANOSTACHYUM* SPECIES IN NEPAL

Leaf sheath ligule up to more than 3mm long .... *falcatum*
Leaf sheath ligule always less than 3mm long:-

        leaf sheath hairy, with prominent and persistent
        auricles, bearing spreading bristles .... *intermedium*

        leaf sheath lightly hairy or glabrous, with
        quickly deciduous or absent auricles .... *khasianum*

note: *Drepanostachyum* species are not well known, and this is a provisional key.

Drepanostachyum falcatum   (Nep. *tite nigalo, diu nigalo*)   T22, T35

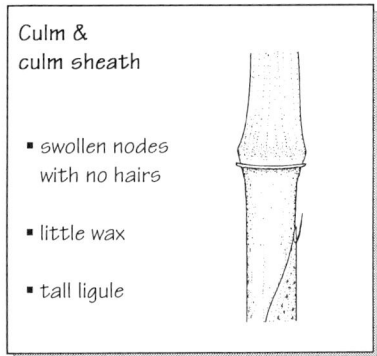

Culm &
culm sheath

- swollen nodes
  with no hairs

- little wax

- tall ligule

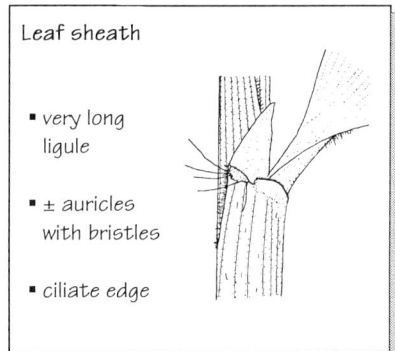

Leaf sheath

- very long
  ligule

- ± auricles
  with bristles

- ciliate edge

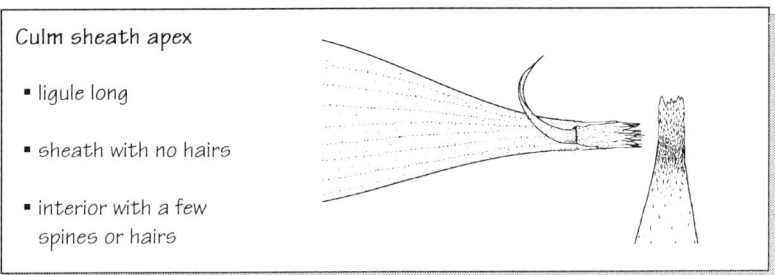

Culm sheath apex

- ligule long

- sheath with no hairs

- interior with a few
  spines or hairs

A variable species, found in drier subtropical forest and also cultivated around farmland. The culms are up to 2cm in diameter and 4m tall when the clumps are cultivated, but they are usually much smaller in the forest because of browsing.

This species can be recognised by the very long leaf sheath ligule. *Borinda chigar* also has very long ligules, but it occurs at higher altitudes, and has much narrower leaves with prominent cross-veins. This species is found from 1,000 to 2,000m in western Nepal. In Kaski District it has small ciliate leaf sheath auricles. In Palpa District it has long cilia on one leaf sheath edge and hairs under the leaves. When cultivated the main use of this species is basket-making, although it also provides useful animal fodder in winter. As with all *Drepanostachyum* species the culms are not very straight, and have rather swollen nodes and many branches, so it is not an ideal species for weaving material. However, it can easily be grown at subtropical altitudes, which makes it a very valuable species.

It is planted in gullies, beside paths, and on waste land but can also be planted on terrace risers, where it is very effective in soil stabilisation. The traditional propagation method is very successful in this species, but a large rhizome section with 2-4 culms should be used. Smaller plants may not survive.

Drepanostachyum intermedium   (Nep. *tite nigalo*)   T1

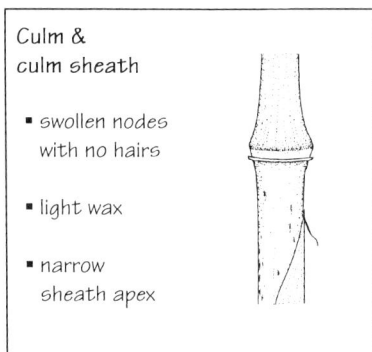

Culm &
culm sheath

- swollen nodes
  with no hairs

- light wax

- narrow
  sheath apex

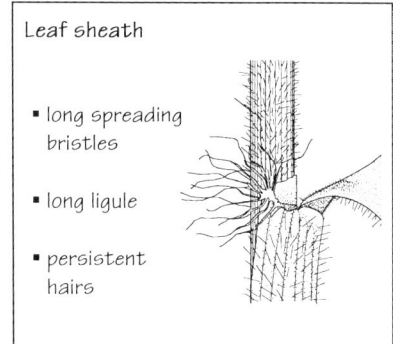

Leaf sheath

- long spreading
  bristles

- long ligule

- persistent
  hairs

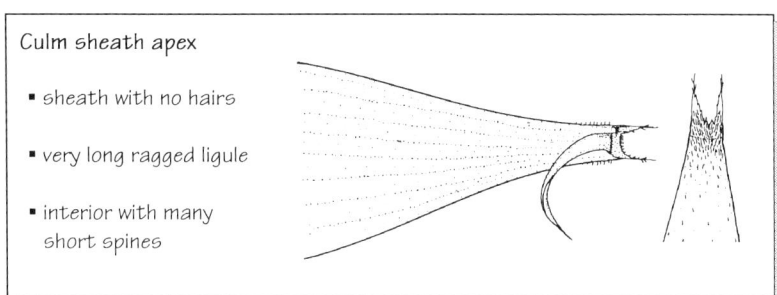

Culm sheath apex

- sheath with no hairs

- very long ragged ligule

- interior with many
  short spines

A species found in evergreen oak and chestnut forest, and also cultivated around subtropical farmland. The culms can be up to 2cm in diameter and 4m tall, but they are usually much smaller in the forest because of browsing.

It can easily be recognised by the well developed and persistent leaf sheath auricles with widely spreading bristles, and by the hairs on the leaf sheaths and the undersides of the leaves, which are very dense in some cultivated plants.

It is found from 1,000m to 2,000m in eastern Nepal. When cultivated, the main use of this species is basket-making, although it also provides useful animal fodder in winter. The culms are not very straight and have rather swollen nodes and many branches, but the ease of propagation at subtropical altitudes makes this a very valuable species.

This is a resilient bamboo which can survive on drier sites than species such as *Himalayacalamus hookerianus,* which could provide much straighter weaving material in moister sites. It is planted in gullies and on waste land but can also be planted on terrace risers, where it is very effective in soil stabilisation.

The traditional propagation method is very effective in this species, as it produces a large number of new shoots at a fast rate, and the rhizomes can be extracted easily.

Drepanostachyum khasianum   (Nep. *ban nigalo*)   T33

Culm & culm sheath	Leaf sheath
• swollen nodes with no hairs • little wax • culm smooth	• long ligule with dense hairs • variable deciduous auricles • few hairs

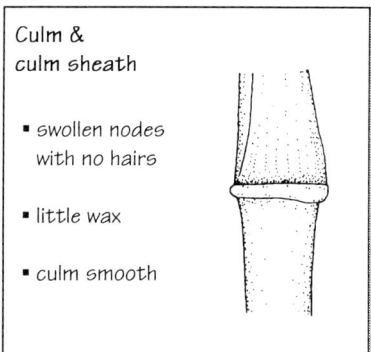
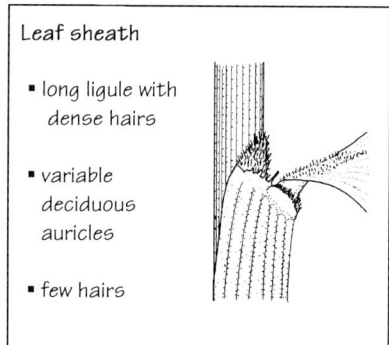

Culm sheath apex

• ligule short

• sheath with no hairs

• interior with a few spines or hairs

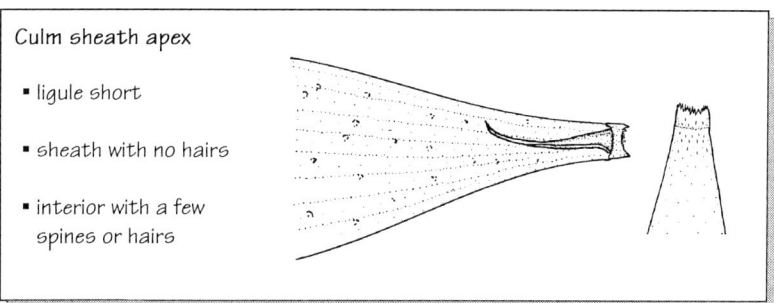

This species has been found in several districts of central Nepal. It is the most common forest bamboo species between 1,000m and 2,000m, and it flowered in Charikot and Nagarkot in 1985. It is only found in the forest and is not cultivated. Where the forest is protected or adequately managed, this bamboo is regularly harvested, but much of its natural habitat has been reduced to degraded scrub, where it cannot reach a large enough size to be of any use.

It is heavily browsed and is usually reduced in size to below 1cm in diameter and 3m in height, although like other *Drepanostachyum* species it could reach 5m in height if it were protected. This species has no prominent characters by which it can be recognised. It is identified by the absence of characters rather than by their presence. The culm sheaths have no hairs, auricles, or bristles, and a short ligule. The leaf sheath has few hairs, a small deciduous auricle, and a short ligule. It can be distinguished from all the species of *Himalayacalamus* by the presence of rough spines inside the culm sheath on and below the ligule.

It is presently identified as *D. khasianum*, a species with similar vegetative characteristics found in Meghalaya, but this is tentative as the flowers of that species are not known.

8. HIMALAYACALAMUS

Clump-forming thornless bamboos up to 8m tall, found from 1,800-2,500m in cool broadleaved forest, and also widely cultivated. They have single flowers, short buds, and 15-40 branches. Leaves do not have the clear cross-veins seen in *Thamnocalamus* and *Borinda* species, and internodes are generally less than 40cm long, shorter than those of species of *Cephalostachyum*. Although similar to species of *Drepanostachyum,* they differ in many ways. Branch buds at mid-culm nodes have fewer initials (compare figs. 35 and 31). Branches are fewer, usually around fifteen in the first year. They vary in size, are more erect, and do not spread right around the culm. The basal culm internodes increase in length progressively. The culm sheaths are completely smooth inside, and are usually broad towards the top, with a short ligule. They are more tolerant of cold and are often found at higher altitude, but they are also less drought tolerant. The new shoots of several species are edible and often harvested.

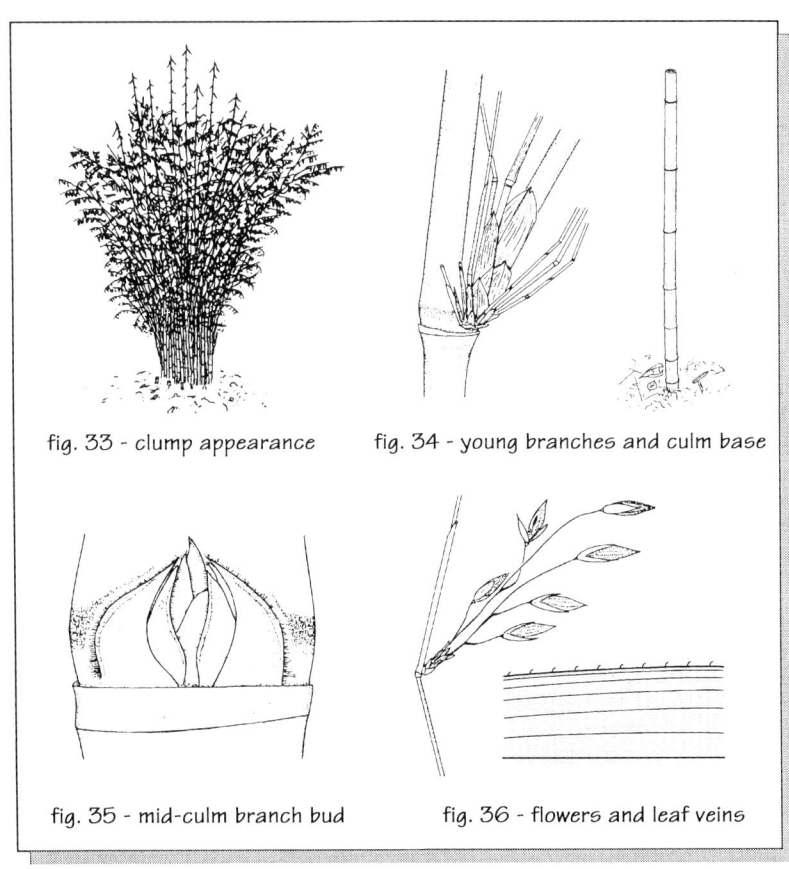

fig. 33 - clump appearance

fig. 34 - young branches and culm base

fig. 35 - mid-culm branch bud

fig. 36 - flowers and leaf veins

KEY TO *HIMALAYACALAMUS* SPECIES IN NEPAL

Culm sheath ligule tall and feathered, culm sheath with ring of hairs
at the base . . . . *fimbriatus*

Culm sheath ligule short and serrated, culm sheath with no hairs
at the base:-

    culm surface finely ridged . . . . *porcatus*
    culm surface smooth:-

        culm sheath very rough, or shortly hairy at the top . . . . *asper*
        culm sheath with no hairs or very slightly rough:-

            culm sheath blades upright:-

                culm sheaths tough, broad at the top; culm sheath
                edges with long copper-coloured cilia; culm
                internodes up to 40cm long . . . . *cupreus*

                culm sheaths thin, narrow at the tpop; culm sheath
                margins without long cilia; culm internodes up
                to 20cm long . . . . *brevinodus*

            culm sheath blades bent backwards:-

                culm sheaths tall, narrow at the top; new culms with
                thick blue wax . . . . *hookerianus*

                culm sheaths short, broad at the top, asymmetrical;
                new culms with thin white wax . . . . *falconeri*

Himalayacalamus asper (Nep. *ghunre nigalo, malinge nigalo*)   T23/T29

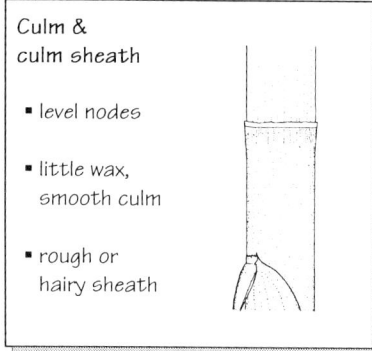

Culm &
culm sheath

- level nodes

- little wax,
  smooth culm

- rough or
  hairy sheath

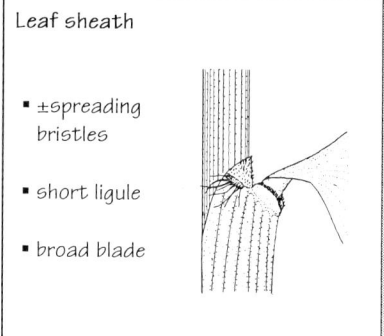

Leaf sheath

- ±spreading
  bristles

- short ligule

- broad blade

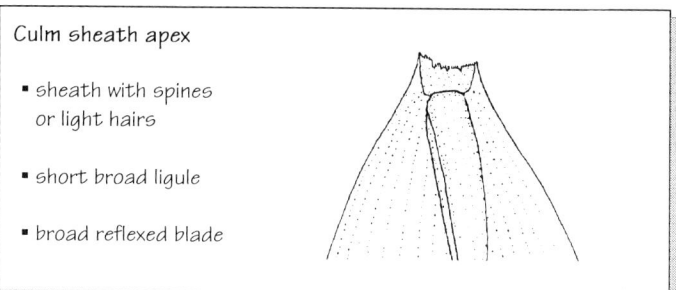

Culm sheath apex

- sheath with spines
  or light hairs

- short broad ligule

- broad reflexed blade

A little-known species found so far in only two localities in western and central Nepal, in temperate forest between 1,800m and 2,300m.

The rough or lightly hairy culm sheaths are the best way of separating it from *Himalayacalamus falconeri*, which has very smooth culm sheaths with no spines or hairs. In addition, the leaves and the culm sheath blades are very broad, and the nodes are very level, not raised as much as those of *H. falconeri*.

In the Langtang Valley, where the culm sheaths are slightly hairy, this species is the best bamboo for weaving, and it is regularly harvested. It is known

there as *malinge nigalo*. It differs from *Himalayacalamus porcatus* which also grows in that area in its smooth culms with no ridges, and the absence of bristles on the culm sheath shoulders.

In the Seti Khola valley of Kaski District this species has rough culm sheaths with tiny points on the back, and spreading bristles on the leaf sheaths. It is found below a belt of the far superior bamboo, *Himalayacalamus cupreus*, which has much longer culm internodes. It is known there as *ghunre nigalo* and is not harvested. It is also cultivated around villages in that area, but probably as a remnant of the natural vegetation.

Himalayacalamus brevinodus (Nep. *malinge nigalo*)   T3/2B

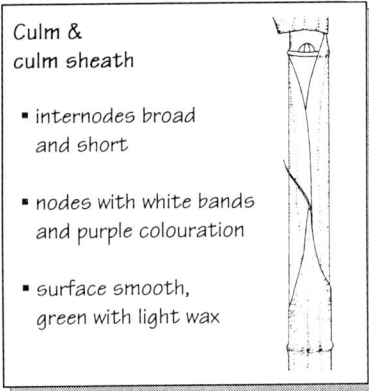

Culm &
culm sheath

- internodes broad
  and short

- nodes with white bands
  and purple colouration

- surface smooth,
  green with light wax

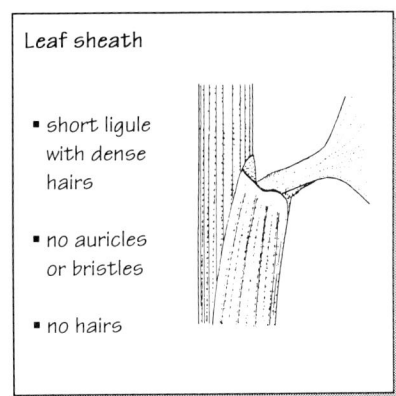

Leaf sheath

- short ligule
  with dense
  hairs

- no auricles
  or bristles

- no hairs

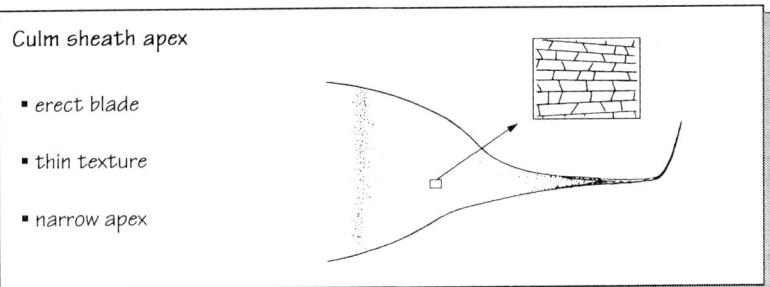

Culm sheath apex

- erect blade

- thin texture

- narrow apex

This valuable bamboo is cultivated between 1,800m and 2,200m in East Nepal, where it reaches a diameter of 2.5cm and a height of 9m.

It is easy to separate this bamboo from *Drepanostachyum intermedium* as it has no hairs or bristles on the culm sheaths. The short broad culm internodes, which never exceed 20cm distinguish it from *Himalayacalamus falconeri*. The culms are greener than those of *Himalayacalamus hookerianus* with a much thinner waxy coating. There are strong bands of purple above each node, and prominent white rings where sheaths have fallen off. Culm sheaths can be similar to those of *H. hookerianus* with long blades and broad or narrow tops, but they are more delicate than those of either *H. falconeri* or *H. hookerianus,* with erect blades.

The culms of this bamboo provide weaving material which is superior to that of *Drepanostachyum intermedium,* although it is probably not as good as that of *Himalayacalamus hookerianus* because of the shorter length of the culm internodes. The shoots are also edible, and the leaves can be used as animal fodder.

This bamboo is easy to propagate and is widely planted around higher villages in eastern districts of Nepal. Its flowers are yet not known.

Himalayacalamus cupreus   (Nep. *malinge nigalo*)   T24

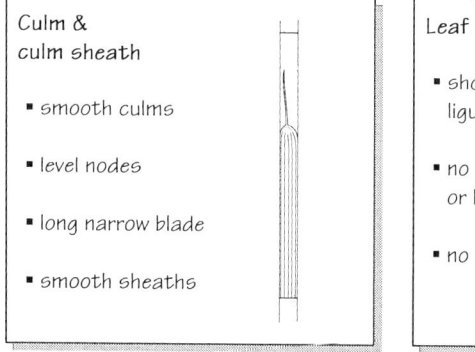

Culm &
culm sheath

- smooth culms

- level nodes

- long narrow blade

- smooth sheaths

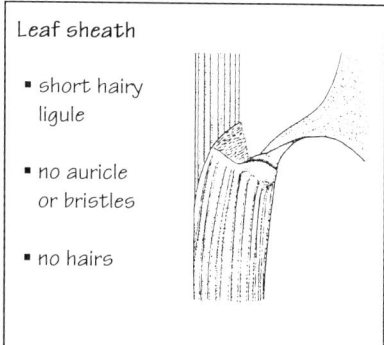

Leaf sheath

- short hairy
  ligule

- no auricle
  or bristles

- no hairs

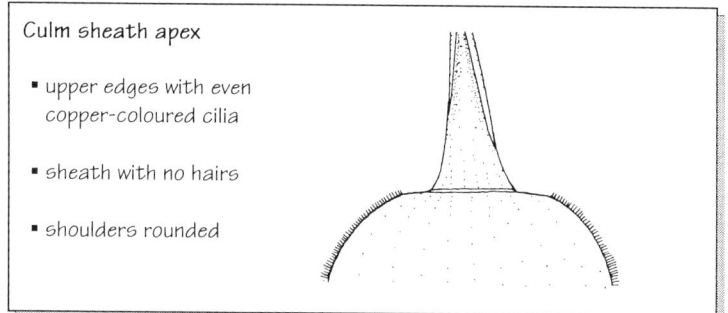

Culm sheath apex

- upper edges with even
  copper-coloured cilia

- sheath with no hairs

- shoulders rounded

This species is found in cool temperate forests in Kaski District, between 2,300m and 2,800m. It is the highest altitude *Himalayacalamus* species found so far, occurring above the range of *Himalayacalamus asper*, but below the truly hardy bamboos in the genera *Borinda* and *Thamnocalamus*.

As a higher altitude species, this bamboo has slight tessellation in its leaf veins, with faint cross-veins which are just visible. The other characteristics which distinguish this species are the extremely long culm internodes, up to 40cm in length, and the prominent copper-coloured cilia along the edges of the long, tough, smooth, culm sheaths. These characteristics clearly separate it from other *Himalayacalamus* species, although its flowers are still not known.

Because of its large culms with long internodes, this is the most sought after bamboo in the area, and it is now carefully managed in order to control its exploitation. The new shoots are very palatable, and they are often collected. This practice is now restricted, as it reduces the number and size of culms which can be harvested later. Porters passing through the forest may have their loads inspected to check that they are not removing shoots of this species.

Himalayacalamus falconeri (Nep. *thudi nigalo, singhane*) T27

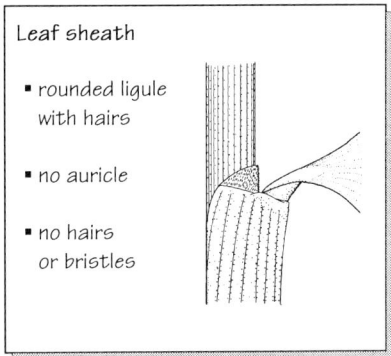

Culm & culm sheath

- swollen nodes with red bands

- broad shoulders

- sheath smooth, often striped

Leaf sheath

- rounded ligule with hairs

- no auricle

- no hairs or bristles

Culm sheath interior

- ligule short and broad

- apex membranous with visible cross-veins

- interior smooth, no hairs or spines

A locally common bamboo in cool broadleaved forests of central and eastern Nepal between 2,000m and 2,500m. It is common at the summit of Phulchowki in the Kathmandu Valley.

This species can be distinguished from other *Himalayacalamus* species by the absence of spines, hairs, or auricles on the bullet-shaped culm sheaths, and the smooth fairly short culm internodes. The culm sheaths have a short broad ligule, and are often striped with yellow and purple lines.

The relatively large size and flexibility of the culms of this species make it a desirable bamboo for weaving, and it is widely harvested. The shoots are edible and they are widely collected from the forest, which can sometimes conflict with use of the older culms for weaving. Small bamboo shoots on sale in Kathmandu markets are usually from this species.

The young shoots have a thick glutinous covering, which leads to the local name in eastern Nepal, *singhane nigalo*. This covering may help to reduce attack by insects such as shoot borers.

Sporadically-flowering clumps are common, but seed has not been produced. Propagation of this species by the traditional technique has been undertaken around higher altitude villages in central and eastern Nepal, but it usually harvested from natural forest stands instead.

Himalayacalamus fimbriatus (Nep. *tite nigalo*)   T21

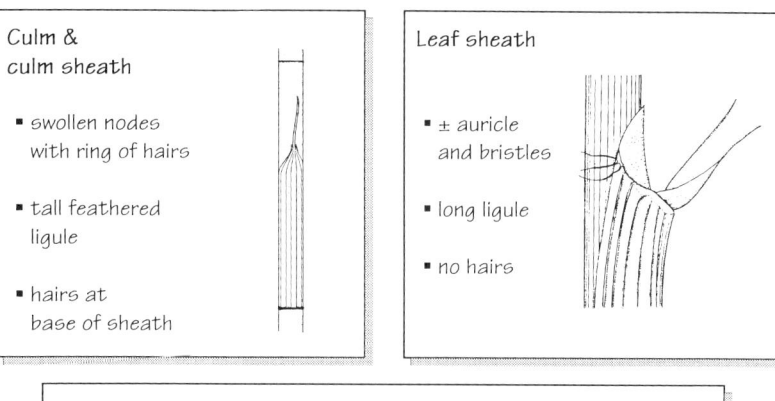

Culm &
culm sheath

- swollen nodes
  with ring of hairs

- tall feathered
  ligule

- hairs at
  base of sheath

Leaf sheath

- ± auricle
  and bristles

- long ligule

- no hairs

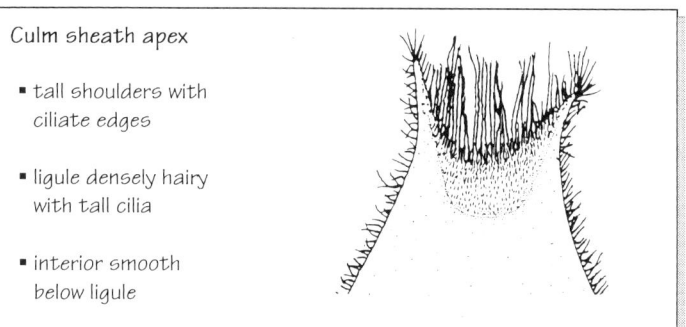

Culm sheath apex

- tall shoulders with
  ciliate edges

- ligule densely hairy
  with tall cilia

- interior smooth
  below ligule

A widely cultivated species of central and western Nepal, common around most villages between 1,100m and 1,800m.

This species has a very distinctive ring of dense orange-brown hairs at the base of the culm sheath, which distinguishes it from all other small bamboos in Nepal. The culm sheath ligule is rough on the inside surface, but as in other *Himalayacalamus* species the sheath below the ligule is completely smooth. The top of the sheath is narrow, as in species of *Drepanostachyum,* and it has a densely hairy, tall, fringed ligule. The leaf sheaths are quite variable, and some

sheaths have no auricles or bristles at all. Other sheaths have small deciduous auricles when they are young. In Chautara the leaf sheaths have larger, more persistent auricles with spreading bristles.

This is a widely planted bamboo in central Nepal, which provides an annual supply of weaving material, as well as animal fodder. It grows well in relatively dry situations in full sunshine. The shoots are very bitter, and are never eaten. It has been cultivated for a very long time, and it is said that it does not ever flower. No flowering specimens have been collected.

Himalayacalamus hookerianus   (Nep. *padang*)   T4

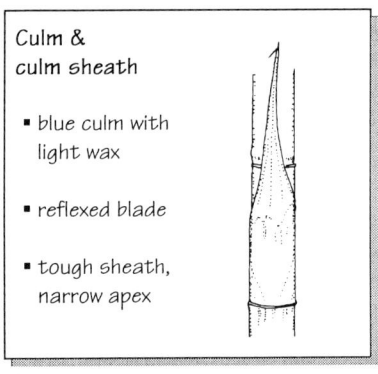

Culm &
culm sheath

- blue culm with
  light wax

- reflexed blade

- tough sheath,
  narrow apex

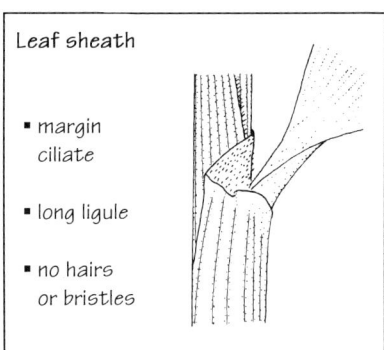

Leaf sheath

- margin
  ciliate

- long ligule

- no hairs
  or bristles

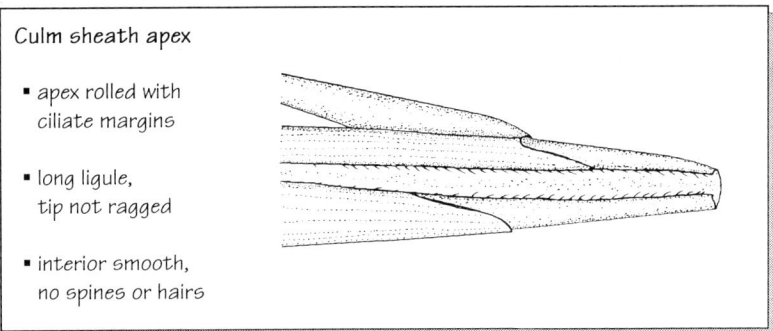

Culm sheath apex

- apex rolled with
  ciliate margins

- long ligule,
  tip not ragged

- interior smooth,
  no spines or hairs

A common cultivated bamboo of East Nepal, found from 2,000m to 2,500m, up to 3cm in diameter and up to 7m tall.

It can easily be recognised by the blue colour of the new culms, and by the long narrow necks of the tough culm sheaths. *Drepanostachyum* species have similar culm sheaths but the interior is rough at the top, and they have greener culms. *Ampelocalamus patellaris* has ridges on the culms and projecting corky collars at culm nodes. *Himalayacalamus brevinodus,* which is also cultivated in East Nepal, has shorter internodes and thinner culm sheaths with erect blades. *Himalayacalamus fimbriatus* has dense hairs at the base of the culm sheaths.

The principle use of this species is basket-making. It can provide animal fodder, but the leaves are small. It produces high quality weaving material. The culms have fewer branches towards the base than *Drepanostachyum* species, and much longer internodes than *Himalayacalamus falconeri* so that splitting them into weavable strips is easier.

It is planted in gulleys, on waste land, and on terrace risers. The lack of branches in the lower half of the culm makes propagation by the traditional technique quite difficult. A longer pole must be used to ensure the successful development of branches from the buds at the top.

Himalayacalamus porcatus (Nep. *seto nigalo*)   T39

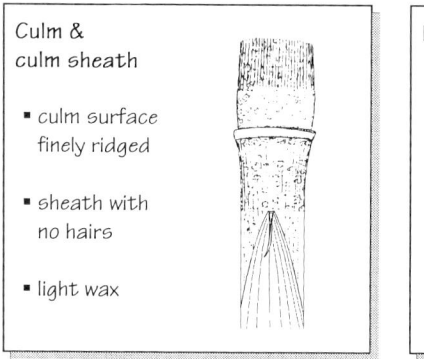

Culm &
culm sheath

- culm surface
  finely ridged

- sheath with
  no hairs

- light wax

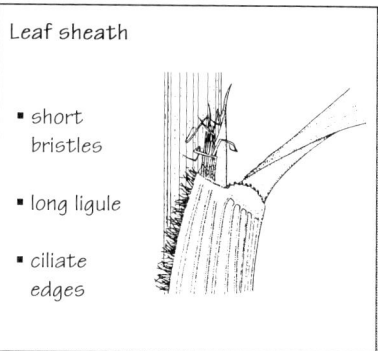

Leaf sheath

- short
  bristles

- long ligule

- ciliate
  edges

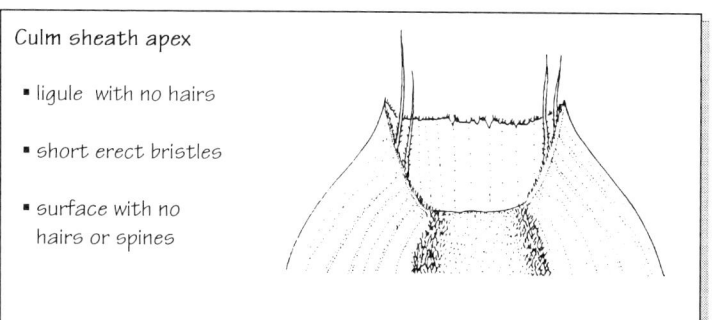

Culm sheath apex

- ligule with no hairs

- short erect bristles

- surface with no
  hairs or spines

This is a rare species from around 2,000m to 2,300m in central Nepal, cultivated, or growing naturally in broadleaved forest, with culms up to 2.5cm in diameter, and up to 6m tall.

The finely-ridged culm internodes of this species distinguish it clearly from all other *Himalayacalamus* species. The ridged culms are similar to those of *Borinda emeryi* of East Nepal, but it has very different buds, more branches, and leaves without any cross-veins. The ridged culms are also similar to those of *Ampelocalamus patellaris,* but this species does not have prominent corky nodes or fringed culm sheath edges. The leaf sheaths have more cilia on the edges than other *Himalayacalamus* species, and upright bristles, but no auricles. The flowers differ from those of other *Himalayacalamus* species such as *H. falconeri* and *H. hookerianus* in being rough, with fine spines on the spikelets.

Although cultivated clumps of this species have been seen, it is not a favoured species for weaving. The culms are brittle, and when split the edges are extremely sharp. It is said that the culm sections can easily cut the hands of those who try to weave baskets or mats from them.

Flowering clumps of this species were found in 1984.

9. MELOCANNA

A spreading thornless tropical bamboo introduced from Bangladesh, with straight upright culms up to 21m tall and 7cm in diameter, arising from rhizomes which are up to 2m long. It is only found below 1,400m. This is the largest spreading bamboo native to the region, highly valued for straightness, durability, and excellent paper-pulp. The large size of the culms is usually sufficient to distinguish it from other spreading bamboos in Nepal. The long rhizomes and well-separated culms can distinguish it from clump-forming bamboos of similar size. The culms are round, without the flattened sides of the introduced spreading Chinese genus *Phyllostachys*, and there are many branches at each node. The culm sheath is very distinctive, with a long narrow blade. Culm buds are short and tough, and closed at the front. There are up to 40 branches from each culm node, and they are all similar in size. Leaves have no cross-veins. One species is known, forming extensive stands in Bangladesh.

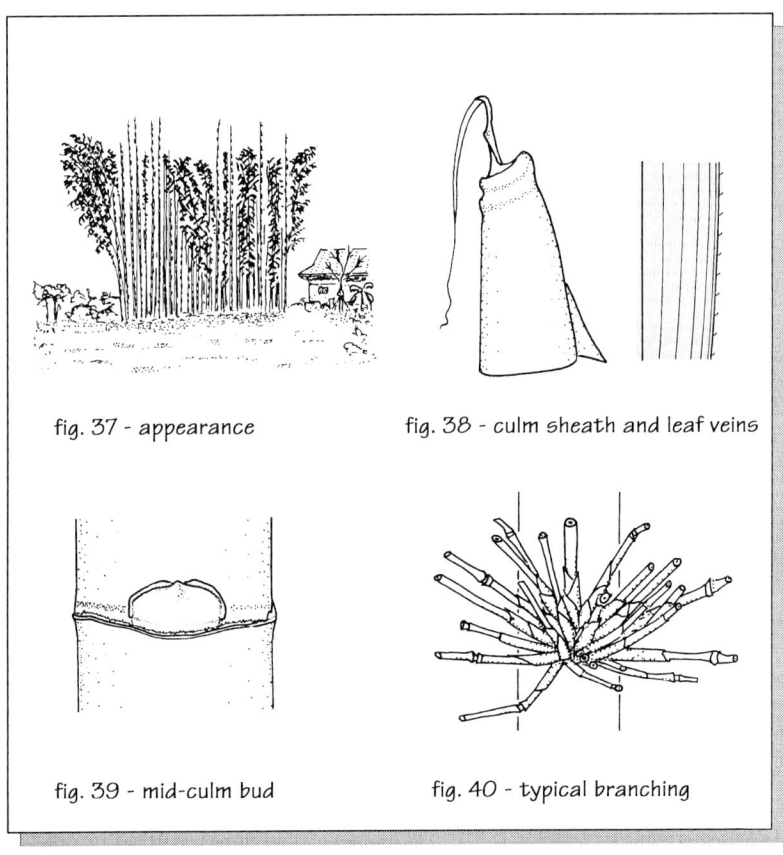

fig. 37 - appearance

fig. 38 - culm sheath and leaf veins

fig. 39 - mid-culm bud

fig. 40 - typical branching

Melocanna baccifera   (Nep. *lahure bans*)   M1

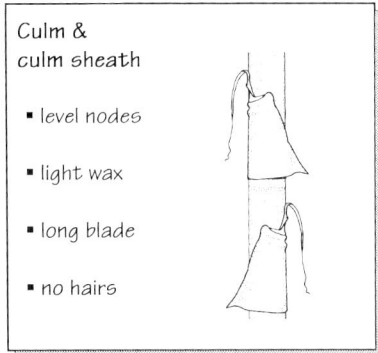

Culm &
culm sheath

- level nodes

- light wax

- long blade

- no hairs

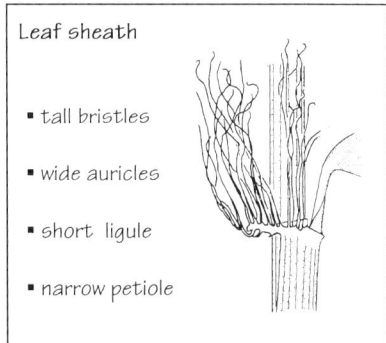

Leaf sheath

- tall bristles

- wide auricles

- short ligule

- narrow petiole

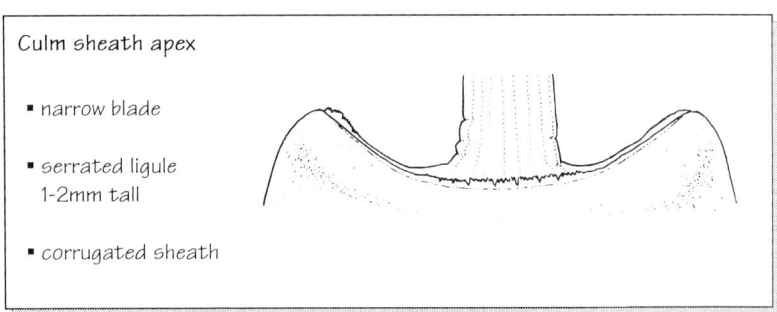

Culm sheath apex

- narrow blade

- serrated ligule
  1-2mm tall

- corrugated sheath

A distinctive bamboo commonly cultivated in the eastern terai and occasional in other areas such as Palpa district. It forms graceful open stands of medium-sized straight upright culms, reaching 12m in height and 5cm in diameter. It requires high temperatures and rainfall of over 2m per year to reach its maximum potential height of 21m.

The culm sheaths are covered in white hairs at first and have two strong waves towards the top. There is a ridge on the outside of the sheath where the blade is attached (callus). In most other bamboos this is normally only seen on leaf sheaths. The blade is sword-shaped and longer than the sheath. The leaf sheath auricles are prominent with very long erect white wavy bristles. The fruits are famous for their large size and shape, similar to that of a pear, and they often germinate before falling off the mother plant, making storage of the seed very difficult.

The culms are smaller than those of *Bambusa* or *Dendrocalamus* species, but are thick-walled (solid at the base), very straight, and said to be termite resistant. They provide a good general purpose construction material, and are also widely used for mats.

This species cannot be propagated by culm cuttings. The traditional planting technique is most appropriate, using a short rhizome length - the long rhizome neck is not required.

10. YUSHANIA

Spreading thornless frost-hardy bamboos, forming dense thickets or covering large areas, with upright culms from 1m to 4m tall, found in temperate forests and open grazing areas, from 1,800m to 3,600m, often stunted by the browsing of livestock. Leaves have clear cross-veins, unlike the leaves of the subtropical spreading species *Melocanna baccifera,* which only has parallel veins. The culms are not prominently ridged as in *Borinda,* and the branches are fewer in number and more upright The young culms of most species are rough below the nodes, while those of *Thamnocalamus* are always smooth. *Yushania* species have rhizomes of more than 30cm with rootless necks. The rhizomes may be solid, or hollow with no dividing walls at their nodes. *Phyllostachys* and *Arundinaria* species have roots all along the rhizome (fig. 38 cf. fig. 54), and their hollow rhizomes are closed at nodes. Larger species can prevent tree regeneration after clear-felling, and can be pernicious weeds.

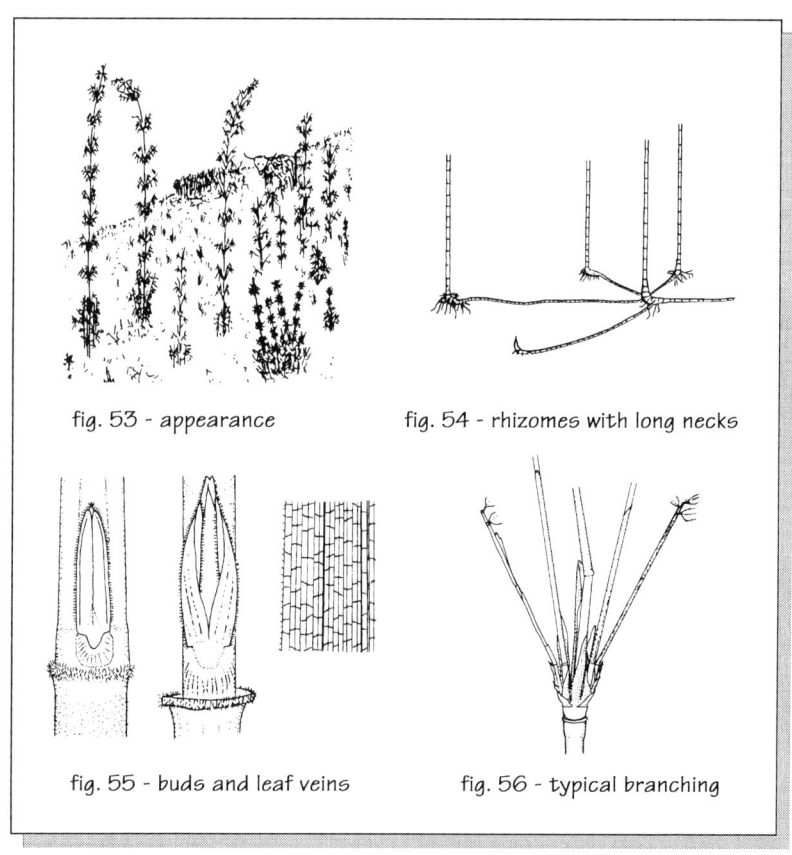

fig. 53 - appearance

fig. 54 - rhizomes with long necks

fig. 55 - buds and leaf veins

fig. 56 - typical branching

KEY TO *YUSHANIA* SPECIES

Rhizome necks solid; leaf edges similar, both thin .... *maling*

Rhizome necks hollow; leaf edges different
with thick clear band along one edge .... *microphylla*

Yushania maling   (Nep. *malingo, maling, khosre malingo*)   T9

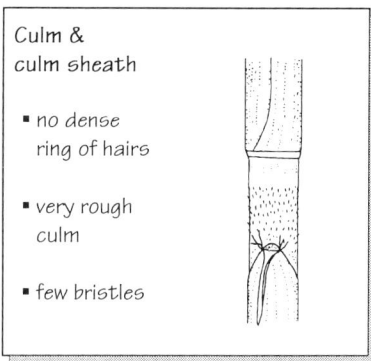

Culm &
culm sheath

- no dense
  ring of hairs

- very rough
  culm

- few bristles

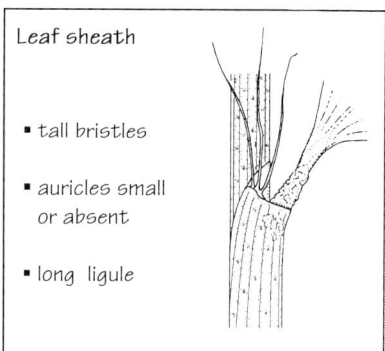

Leaf sheath

- tall bristles

- auricles small
  or absent

- long ligule

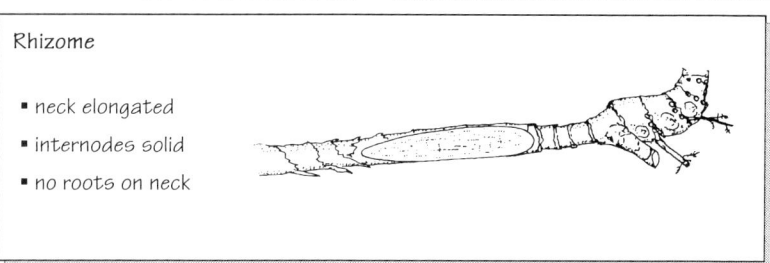

Rhizome

- neck elongated
- internodes solid
- no roots on neck

This species is the most common spreading temperate bamboo in East Nepal. It occurs from 1,600m to about 3,000m.

It is similar to *Thamnocalamus spathiflorus* and *Arundinaria racemosa* but can easily be distinguished from those species by the roughness of the internodes on new culms. The solid rhizome necks and lack of a clear thickened band on either leaf edge distinguish it from the rarer species *Y. microphylla,* which has smoother culms with much more white or black wax below the nodes.

When growing vigorously larger culms may be used for fencing, or sometimes woven into baskets, but they are usually too small for these uses, so that the culms can only be used for making brushes and straws.

This species is normally similar in size to *Y. microphylla,* with culms rarely reaching 3m in height or 1cm in diameter, but when the stands are unusually dense or tall, they can interfere with tree regeneration.

This species was confused with *Arundinaria racemosa* for a long time. It can easily be distinguished from that species by the rough internodes of young culms, and by the long rhizome necks with no roots. The Nepali local names, *maling* or *malingo,* may also be used for several other spreading and clump-forming bamboos.

Yushania microphylla   (Nep. *maling, malingo*)   T45

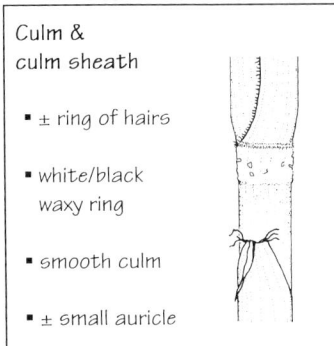

Culm &
culm sheath

- ± ring of hairs

- white/black
  waxy ring

- smooth culm

- ± small auricle

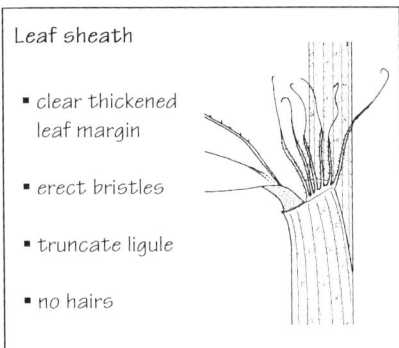

Leaf sheath

- clear thickened
  leaf margin

- erect bristles

- truncate ligule

- no hairs

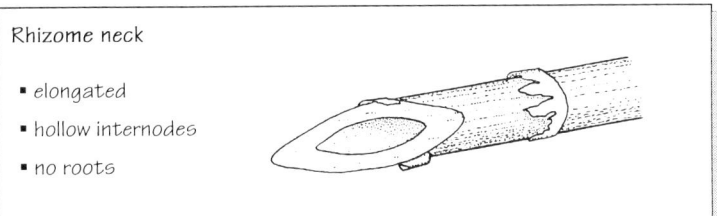

Rhizome neck

- elongated

- hollow internodes

- no roots

This is a rare bamboo of cool temperate areas in Central and East Nepal, between 2,300m and 3,500m. It forms large areas of yak-grazing pasture in Tibet and Bhutan. It is usually found on gently sloping wet areas, rather than steep slopes or gullies.

It is often heavily browsed and also often burnt, so that it is commonly less than 1m tall, often with balls of short branches at each node and leaves less than 3cm long. The culms can be up to 3m tall and 1.5cm in diameter, with leaves of up to 10cm, when it is protected from grazing animals.

This species can be distinguished from other *Yushania* species and from *Arundinaria racemosa* by the thick transparent band along one edge of its leaves. It also has a persistent flaky ring of wax below the culm nodes, which turns from white to black with age. In addition the rhizome necks are hollow, even at the nodes, producing long soft hollow cylinders. It is a variable species, and some plants are more hairy than others, sometimes with a ring of hairs at the culm sheath base. New culms can be slightly rough or smooth, and leaf sheath auricles may be absent or pronounced.

This bamboo is usually too short to shade out tree regeneration, and is important for livestock and wildlife in the winter months, both in open grazing areas and in the forest. The long hollow rhizome necks may assist in drainage and aeration in waterlogged sites.

11. ARUNDINARIA

Spreading thornless frost-hardy bamboos, with upright round culms up to 3m tall, rhizomes with roots at all nodes, and simple branching. They are found in temperate forest and grazing areas, from 2,900 to 3,600m, often mixed with *Yushania* or *Thamnocalamus* species. Similar in appearance to small *Yushania* species, but with culms which are always smooth and have little or no wax. The rhizomes are also different in that they continue under the ground indefinitely, with roots at all nodes and culms branching upwards at intervals. *Chimonobambusa* from West Bengal has very similar rhizomes, but its culm nodes are raised and often bear a ring of sharp thorns. *Arundinaria* buds are tall, similar to those of *Yushania* and *Borinda*, but the branching is simpler, with a single branch leaving the culm, but then branching repeatedly in a fan-shaped arrangement. The leaves have very prominent cross-veins, unlike those of the spreading bamboo from subtropical areas, *Melocanna baccifera*.

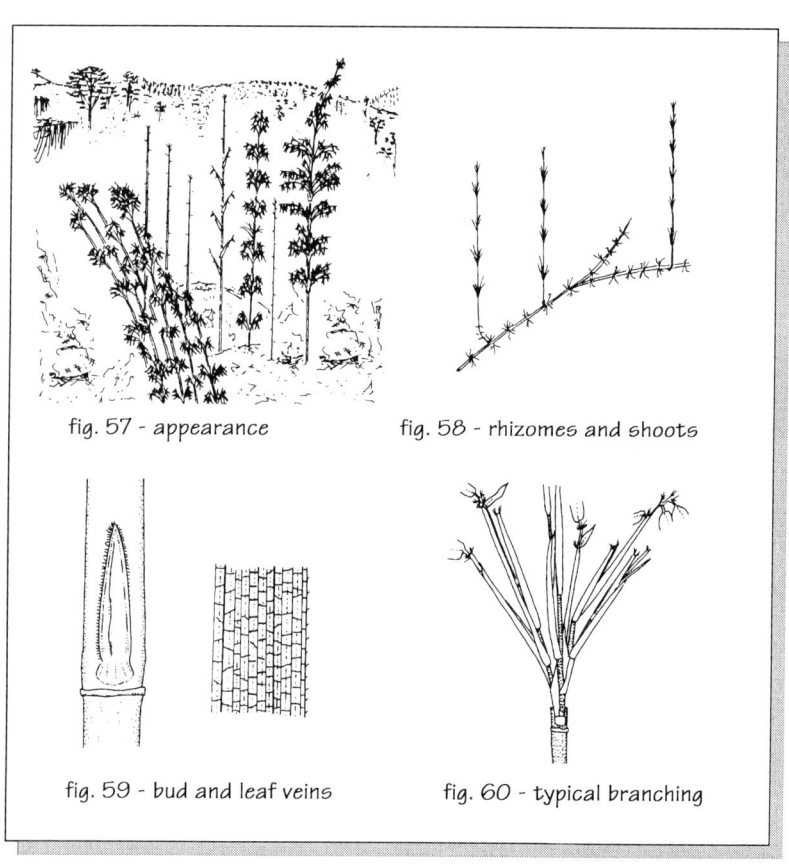

fig. 57 - appearance

fig. 58 - rhizomes and shoots

fig. 59 - bud and leaf veins

fig. 60 - typical branching

Arundinaria racemosa

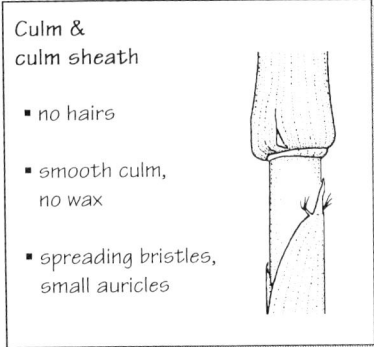

Culm & culm sheath

- no hairs

- smooth culm, no wax

- spreading bristles, small auricles

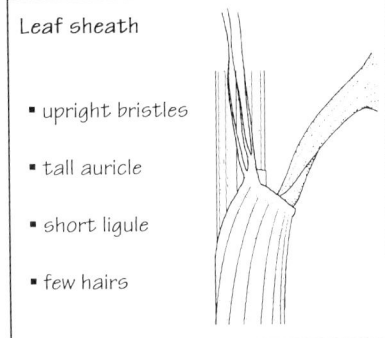

Leaf sheath

- upright bristles

- tall auricle

- short ligule

- few hairs

Rhizome

- internodes all elongated & hollow

- nodes all solid, & all bearing roots

This is a rare bamboo, likely to be found only in high altitude coniferous forest of East Nepal above 2,900m. It forms a component of open bamboo pastureland in Bhutan. It is usually found in better drained or more sloping sites than *Yushania* species, but not usually on the steepest sites, where *Thamnocalamus spathiflorus* thrives.

It is usually less than 2m tall and 1cm in diameter, with leaves up to 10cm long, but like *Yushania* species it is often stunted by grazing and burning. Larger plants may be found in more protected sites.

This species is best distinguished from *Yushania* species by the totally different form of rhizome with roots at all nodes.

It can be distinguished quickly from *Yushania microphylla*, the common spreading species with which it is often found, by the absence of a thick transparent band along one edge of its leaves. It also has fewer, stronger, less scabrous bristles on the leaf sheath auricles. New culms are always smooth, without any roughness or wax below the nodes.

This species is usually too small to shade out tree regeneration, and tends to form more open stands than *Yushania* species. It may be important for grazing of livestock and for wildlife, and is also used for making arrows, brushes and drinking straws.

CHECKLIST OF SPECIES AND AUTHORITIES
(with synonyms in italics)

Ampelocalamus Chen, Wen & Sheng

    A. patellaris (Gamble) Stapleton

        Dendrocalamus patellaris Gamble

        Chimonobambusa jainiana Das & Pal

        Drepanostachyum jainianum (Das & Pal) R.B. Majumdar

Arundinaria Michaux

    A. racemosa Munro

        Fargesia racemosa (Munro) Yi

        Yushania racemosa (Munro) R.B. Majumdar

Bambusa Schreber

    B. alamii Stapleton

    B. balcooa Roxburgh

        Dendrocalamus balcooa (Roxburgh) Voigt

    B. multiplex (Lour.) Raeusch. ex J.A. & J.H. Schult.

        Bambusa glaucescens (Willd.) Holttum

        Bambusa nana Roxb.

        Arundo multiplex Lour.

        Ludolfia glaucescens Willd.

    B. nepalensis Stapleton

    B. nutans Wallich ex Munro **subsp. cupulata** Stapleton

        Bambusa macala Wallich

    B. nutans Wallich ex Munro **subsp. nutans**

    B. tulda Roxburgh

        Dendrocalamus tulda (Roxburgh) Voigt

    B. vulgaris Schrader ex Wendland

Borinda Stapleton

    B. emeryi Stapleton

Cephalostachyum Munro

    C. latifolium Munro

        Schizostachyum latifolium (Munro) R.B. Majumdar

Cephalostachyum fuchsianum Gamble

Schizostachyum fuchsianum (Gamble) R.B. Majumdar

Dendrocalamus Nees

D. giganteus Munro

D. hamiltonii Munro **var hamiltonii**

D. hamiltonii Munro **var. undulatus** Stapleton

D. hookeri Munro

D. strictus (Roxb.) Nees

Bambos stricta Roxb.

Bambusa stricta (Roxb.) Roxb.

Drepanostachyum Keng f.

D. falcatum (Munro) Keng f.

Arundinaria falcata Munro

Chimonobambusa falcata (Munro) Nakai

Sinarundinaria falcata (Munro) Chao & Renvoize

D. intermedium (Munro) Keng f.

Arundinaria intermedia Munro

Chimonobambusa intermedia (Munro) Nakai

Sinarundinaria intermedia (Munro) Chao & Renvoize

D. khasianum (Nees) Keng

Arundinaria khasiana Munro

Chimonobambusa khasiana (Munro) Nakai

Himalayacalamus Keng f.

H. asper Stapleton

H. brevinodus Stapleton

H. cupreus Stapleton

H. falconeri (Munro) Keng f.

Thamnocalamus falconeri Munro

Arundinaria falconeri (Munro) Benth. & Hook. f

Drepanostachyum falconeri (Munro) McClintock

Fargesia collaris Yi

Fargesia gyirongensis Yi

H. fimbriatus Stapleton

H. hookerianus (Munro) Stapleton
Arundinaria hookeriana Munro
Sinarundinaria hookeriana (Munro) Chao & Renvoize
Chimonobambusa hookeriana (Munro) Nakai
Drepanostachyum hookerianum (Munro) Keng f.

Melocanna Trinius

M. baccifera (Roxburgh) Kurz
Melocanna bambusoides Trin.

Thamnocalamus Munro

T. spathiflorus (Trin.) Munro **subsp. spathiflorus**
Arundinaria spathiflora Trinius
Arundinaria aristata Gamble
Thamnocalamus aristatus (Gamble) E.G. Camus
T. spathiflorus subsp. *aristatus* (Gamble) McClintock

T. spathiflorus (Trin.) Munro **subsp. nepalensis** Stapleton

T. spathiflorus (Trin.) Munro **var. crassinodus** (Yi) Stapleton
Fargesia crassinoda Yi

Yushania Keng f.

Y. maling (Gamble) R.B. Majumdar
Arundinaria maling Gamble
Sinarundinaria maling (Gamble) Chao & Renvoize

Y. microphylla (Munro) R.B. Majumdar
Arundinaria microphylla Munro
Sinarundinaria microphylla (Munro) Chao & Renvoize

GLOSSARY

Technical terms

aerial root	a root growing above the ground, in the air
auricle	an ear-like projection at the top of a sheath, fig. 1
blade	a leaf or equivalent section at the top of a culm sheath, fig. 1
callus	small flaps at top of leaf sheath below petiole
chevron	pattern of V-shaped stripes
cilia	hairs along an edge
ciliate	with hairs along the edge
clump	a collection of many culms growing close together
cross-veins	short veins running across the leaf seen when looking through a leaf held up to the light
culm	the stem or stalk of a grass plant, a pole in large bamboos
dbh	culm diameter measured 1.3m above the ground (breast height)
genus	a group of similar species with the same generic name e.g. *Bambusa*
initials	small parts of a bud which will grow into separate branches
internode	the section of a culm between two nodes
ligule	a projecting tongue where sheath and blade meet, fig. 1
long veins	veins running along the length of the leaf
node	ring around the culm joints where the sheath is attached
petiole	narrow neck between leaf blade and leaf sheath
pulvinus	swelling at base of petiole turning blade to the light
reflexed	bent backwards at more than 90°
rhizome	horizontal underground stem producing roots and new shoots
serrated	like the edge of a saw
scabrous	surface rough to touch with small sharp points
species	a group of similar plants called by the same species name e.g. *strictus*
spreading	not growing in clumps
subspecies	division of a species covering a large geographical area
truncate	straight as though cut off
variety	division of a species found in a small geographical area

Language abbreviations

Nep.	Nepali
Mait.	Maithili
Eng.	English

BIBLIOGRAPHY

CHAO, C. S. & RENVOIZE, S. A. (1989). A revision of the species described under Arundinaria (Gramineae) in Southeast Asia and Africa. *Kew Bull.* 44(2): 349–367.

GAMBLE, J. S. (1896). The Bambuseae of British India. *Ann. Roy. Bot. Gard. (Calcutta)* 7(1):1–133

JACKSON, J.K. (1987). *Manual of Afforestation in Nepal.* Forestry Research Project, Kathmandu.

KENG, P. C. (1982–3). A revision of genera of bamboos from the world. *J. Bamboo Res.* 1(1):1–19; 1(2):31–46; 2(1):11–27; 2(2):1–17.

MAJUMDER, R. B. (1989). In Karthikeyan et al., *Flora Indicae, Enumeratio Monocotyledonae.* 274-283. Botanical Survey of India, Howrah, Calcutta.

McCLURE, F. A. (1966). *The bamboos: a fresh perspective.* Harvard University Press, Cambridge, Mass.

___ (1973). Genera of bamboos native to the new world. *Smithsonian Contr. Bot.* 9: 1–148.

MUNRO, W. (1868). A monograph of the Bambusaceae. *Trans. Linn. Soc. London* 26:1–157.

NAPIER, I. & ROBBINS, M. (1989). *Forest seed and nursery practice in Nepal.* Forestry Research Project, Kathmandu.

SODERSTROM, T. R. & ELLIS, R. P. (1987). The position of bamboo genera and allies in a system of grass classification. In Soderstrom et al.(eds.). *Grass Systematics and Evolution:* 225-238. Smithsonian Institution Press.

STAPLETON, C. M. A. (1991). A morphological investigation of some Himalayan bamboos with an enumeration of taxa in Nepal and Bhutan. Unpublished PhD thesis, University of Aberdeen.

STAPLETON, C. M. A. (1994a). The bamboos of Nepal and Bhutan Part I: *Bambusa, Dendrocalamus, Melocanna, Cephalostachyum, Teinostachyum,* and *Pseudostachyum* (Gramineae: Poaceae, Bambusoideae). *Edinb. J. Bot.* 51(1): 1-32

STAPLETON, C. M. A. (1994b). The bamboos of Nepal and Bhutan Part II: *Arundinaria, Thamnocalamus, Borinda,* and *Yushania* (Gramineae: Poaceae, Bambusoideae). *Edinb. J. Bot.* 51(2)

STAPLETON, C. M. A. (1994c). The bamboos of Nepal and Bhutan Part III: *Drepanostachyum, Himalayacalamus, Ampelocalamus, Neomicrocalamus,* and *Chimonobambusa* (Gramineae: Poaceae, Bambusoideae). *Edinb. J. Bot.* 51(2)

INDEX